MECHANIZED TUBE WELL INSTALLATION

TOP: WELL-DRIVER PUP IN OPERATIONAL POSITION.
BOTTOM: WELL-DRIVER PUP.

MECHANIZED TUBE WELL INSTALLATION

ROBERT M. STWALLEY III

GRACE L. BALDWIN KAN-UGE

ROGER TORMOEHLEN

PURDUE UNIVERSITY PRESS
WEST LAFAYETTE, INDIANA

Cataloging-in-Publication Data on file at the Library Congress.

978-1-62671-311-6 (paperback)

978-1-62671-312-3 (epdf)

Cover images: Extension drilling rods in front of old water pump truck that extracts water from a hole in the ground at sunny summer day: z1b/iStock via Getty Images Plus. Well-Driver PUP: Baldwin Kan-uge et al., 2023, courtesy of Robert M. Stwalley.

To those in need of convenient access to freshwater across the planet and those who choose to provide it.

CONTENTS

PREFACE

Civilization and economic development begin with access to clean, fresh, drinkable water. Sadly, the United Nations currently estimates that slightly over one in six people on the planet lack sufficient access to this basic human need. Each of us needs a minimum of 20 L of freshwater per day, and this does not include the amount required for bathing, sanitation, and cleaning kitchenware and clothing. Studies clearly show that the burden of daily water retrieval falls disproportionately on women, children, and the elderly and that the amount of water returned to the home falls dramatically with distance to the nearest water source. Even more upsetting is the fact that there is freshwater less than 15 m away for these water access–challenged peoples in underground aquifers. However, the reality is that modern well drilling techniques are too expensive for the majority of these communities to use in accessing this freshwater. Researchers at Purdue University sought to overcome this roadblock by creating a means to mechanize the installation of tube wells. The following collection details the evolution of this technology and should provide enough information to duplicate the developed specialized equipment and process for the installation of these wells. Specific care was taken during the engineering development to minimize the need for complex machinery and costly supplies. The equipment is both operable and maintainable without a significant investment in tools or specialized education. This technology could potentially represent a paradigm shift in how people lacking direct access to sources of water change and upgrade their communities, taking a first major step toward broader economic development. The collection presented in this book is intended to offer guidance for those working to provide low-cost tube well installations and a collegiate-level textbook for specialized engineering design in the environmental and natural resources engineering realm.

The educational objectives are

- To be able to understand subsurface hydrology and the fundamentals of general water well technology,
- To gain a specific understanding of the principles involved in tube well installation and operation,
- To recognize and be able to apply the basic theoretical and mechanical elements of machine technology to the process of driving a tube well, and
- To gain an understanding of the various circumstances that can affect the installation of a tube well.

ACKNOWLEDGMENTS AND COPYRIGHT INFORMATION

The efforts of the tube well mechanization project at Purdue University by Mr. Zackariah J. Horn and Mr. Tyler J. McPheron are gratefully acknowledged for their overall work on the overarching project and original publications. The original publishers of the works contained in this volume, the American Society for Engineering Education (ASEE) and IntechOpen, Limited, are thanked for their efforts in the original dissemination of this work and for their permission to reprint these papers in this collection. The Purdue Agricultural and Biological Engineering Department has also graciously helped provide backing for this project. An artificial intelligence–powered large language model software, ChatGPT, by Microsoft, Incorporated (Redmond, Washington), was used to aid in the organization of some of the original papers, but all manuscript elements are original from the authors. Any errors are entirely the result of the authors. The mention of trade names and commercial products in this article is solely for the purpose of providing specific technical information and does not imply recommendation or endorsement by Purdue University. The findings and conclusions in this publication are those of the authors and should not be construed to represent any official Purdue University determination or policy. Purdue University is an equal opportunity/equal access organization.

The inner cover photographs are by Grace L. Baldwin Kan-uge, who is acknowledged for her artwork and has granted her permission for their use. Permission for the use of all other photographs, drawings, and artwork in the original articles and subsequent republications was granted at the time of original publication by the creators and copyright holders.

The copyright for "Well-Driver PUP," originally presented at the American Society of Agricultural and Biological Engineering's (ASABE) 2018 annual international meeting, was retained by the authors: Zackariah J. Horn and Robert M. Stwalley III. Corresponding author Stwalley has executed a permission statement for republication in this collection, and the ASABE is graciously acknowledged for its original publication of this work.

The copyright for "A Low-Cost Mechanized Tube Well Installer," originally presented at the ASABE's 2019 annual international meeting, was retained by the authors: Zackariah J. Horn and Robert M. Stwalley III. Corresponding author Stwalley has executed a permission statement for republication in this collection, and the ASABE is graciously acknowledged for its original publication of this work.

The copyright for "Improving the Driving Capabilities of a Well-Driver PUP (Purdue Utility Project) to Install Low-Cost Driven Water Wells," originally presented at the ASABE's 2022 annual international meeting, was retained by the authors: Grace L. Baldwin, Tyler J. McPheron, and Robert M. Stwalley III. Corresponding author Stwalley has executed a permission statement for republication in this collection, and the ASABE is graciously acknowledged for its original publication of this work.

INTRODUCTON

The United Nations recognizes 17 goals for sustainable living. Freshwater access is a part of what are known as the WASH objectives: water, sanitation, and hygiene. The lack of freshwater hinders economic development significantly. The overall goal of the broader work collected in this textbook has been to reduce the barriers to freshwater access for developing communities through the use of simple shallow tube well installations. Tube wells differ from traditional drilled wells in the manner of their creation. Drilled wells remove the earthen spoils from the borehole prior to the installation of a casing within the upper levels of the well. Tube wells compress and push aside the soil as the casing is pushed deeper into the ground. Traditionally, tube well installations have been performed manually by lifting and dropping weights or through the use of hammers. The Purdue process outlined in this collection mechanizes the installation, lessening the labor requirement and accelerating the pace of the work. The well driver was attached to a Purdue Utility Platform (PUP), which is an established design for a small off-road vehicle assembled from local parts and requiring minimal manufactured components. This choice was intended to demonstrate that the technology could be applied using minimal advanced components, but the well driver technology could be attached to multiple types of mobile utility vehicles. This overview provides insight into the necessary improvements made to the design as the research progressed, an analytical and experimental study of the potential within the technology, a novel well pipe coupling system developed in parallel with the driving technology, and a general overview of well technology. These topics are intended to serve as a primer for those wishing to duplicate the technology in a developing country and as an advanced engineering design textbook for seniors and graduate students in the agricultural, environmental, and natural resources engineering disciplines.

CHAPTER 1. MATING A WELL DRIVER TO A PUP

The first article describes the initial step in the creation of a system to mechanize tube well installations, which is the mating of a post driver mechanism to a transport vehicle. The vehicle must be capable of off-road work, have an onboard hydraulic system sufficient to run the post driver, and be stable on its suspension. A manufactured hydraulic post driver mechanism is recommended for modification into the tube well installation device. A post driver consists of a ram and guiding frame. The ram is elevated using a one-way hydraulic cylinder, tensioning a spring mechanism that adds to the gravitational potential of the ram when released. Most commercially available units have three-point hitch attachments for use with agricultural tractors. This is a convenient means of coupling the device to the vehicle for operation, but it is not well balanced for transport between sites. The initial Purdue team chose to add a hinge mechanism on the lower link arms and fold the post driver forward over the body of the Purdue Utility Platform vehicle using another hydraulic cylinder to better balance the vehicle for transportation. Finally, the mechanism provided by the manufacturer for securing fence posts during the driving process was modified to accept 2 *in* nominal galvanized pipe. This article additionally describes the proposed proof-of-concept testing program for the combined well driver vehicle as well as the motivation, implications, and potential economic impact of the mechanized tube well driving system. Th article was originally presented at the 2018 annual international meeting of the American Society of Agricultural and Biological Engineers. The poster presented at the meeting is also included.

WELL-DRIVER PUP

ZACKARIAH J. HORN (PURDUE UNIVERSITY AGRICULTURAL &
BIOLOGICAL ENGINEERING) AND ROBERT M. STWALLEY III

ABSTRACT

People living in developing countries or undeveloped regions often do not have proper access to quantities of safe, clean water to fulfill their daily needs. Members of the families, often women and children, walk miles every day to collect surface waters that are frequently contaminated. To improve water availability and quality, a sustainable mechanical solution to access groundwater is in development. A well-driving attachment for a Purdue Utility Platform (PUP) vehicle is proposed to provide a low-cost means for installing tube-type wells in areas of high to medium water table heights. PUP vehicles have a niche in developing countries, as they offer impressive value and utility in comparison to other powered machines. The vehicles are built and sourced using locally available materials with basic tooling. A hydraulic post driver has been attached to the rear of a PUP frame to serve as an impact mechanism, driving a well point and a series of interconnecting pipes to serve as a casing. Successful implementation involves converting a post driver into a reliable, easy-to-operate machine attachment. The main challenges involve analyzing vehicle integrity with high impacts and overall vehicle stability and creating compatibility with a pipe driving system. Modifications must be made to the post driving system to prevent pipe buckling and to guarantee a plumb well drive. Results from testing will give meaningful insight into the viability of the prototype well-driving platform and its reliability in the field. Issues discovered in the development of the unit thus far will be discussed.

KEYWORDS: development, PUP, sand point well, water, water security, well

1. INTRODUCTION

1.1. THE PROBLEM

Most developing areas have an urgent need for clean water. Often due to limited infrastructure and funding, developing areas have no water distribution network. In these areas it is common for women and children to walk miles at a time, retrieving daily what water they can carry for their families. Sources of water, either from the surface or from hand-dug wells, are often contaminated due to limited sanitation practices. Water can carry pathogens such as E. coli, salmonella, and cholera. The diseases from these contaminants are potentially life-threatening if not treated. As the Environmental Literacy Council states, "Simply put, most water industries within developing nations are incapable of supplying adequate and clean water to all citizens" (Environmental Literacy Council, 2015).

1.2. CURRENT SOLUTIONS

Water sourcing methods in developing areas range from primitive to state of the art in process. Hand-dug wells are some of the most common due to the minimal if any cost input. Individual or multiple workers dig deep into the earth until the water table is reached. Often a bucket on a rope is used to retrieve water from the bottom of the well. Although effective, hand-dug wells are very dangerous and unprotected water sources. These wells can be deadly to construct, due to reduced oxygen supply and the risk of wall cave-in. Furthermore, the well is unprotected from contaminants. Contaminants may enter the well by surface runoff directly into the well and by leaching from the lack of a well casing. They are a health hazard in construction and likely to be a hazard in use if contaminated.

Chlorination is trending in development settings due to its ease of use and effectiveness. Upon beginning a chlorine treatment regimen, the results are almost instant. Chlorine, either in liquid or tabular form, is added in a prescribed amount to a quantity of water to be disinfected. Large quantities of water may be treated quickly for drinking and household use. However, chlorine disinfection has drawbacks as well. Chlorine must be delivered on a scheduled basis. If a merchant misses a delivery or is not available for purchase, the water supply is unprotected. Chlorine disinfection is not a sustainable solution.

Professional well drilling is a rarity in these locations although a great solution. Like other methods, there are drawbacks. Well drilling is perhaps the most expensive answer for large quantities of clean, sustainable water. For most development areas, the funding to drill a well is simply not there. Furthermore, well drilling requires experienced labor and specialized equipment. Due to these restrictions, drilling a well is unrealistic for many communities.

1.3. NICHE OF A WELL DRIVER

There exists a niche market for a water sourcing process that is sustainable and cost-effective and requires minimal skilled labor. A driven well is composed of a well point and lengths of galvanized pipe to form the well casing. This type of well is better protected than a hand-dug well while remaining affordable for areas of implementation. Construction of the well is a onetime purchase, requiring very little cost input over its lifespan. Given a method to power this type of well, installation does not require skilled labor and boasts a quick turnaround time.

2. PUP WELL-DRIVING ATTACHMENT

2.1. PUP

The Purdue Utility Platform (PUP) is low-cost utility vehicle designed for developing areas. A PUP is built from all locally sourced materials, such as angle iron and used automotive parts. Furthermore, PUPs are built using limited power and hand tools. Construction is straightforward, with a metal saw, a welder, and a drill. As shown in Figure 1.1, the PUP vehicle frame is triangular, limiting frame torsion off-road and reducing the complexity of the build. Units weigh in at approximately 1,100–1,200 lb_f (500 kg). To date, attachments built to provide utility to PUPs include water pumps, maize grinders, a planter, and even tillage implements. The platform is designed to easily accept new attachments to better benefit the needs of the direct users, improving their quality of life.

FIGURE 1.1. 20IX era PUP vehicles (Purdue University, 2018)

2.2. DESCRIPTION

The heart of the well-driver attachment is a modified Shaver HD-8 hydraulic post driver. At full stroke, the HD-8 produces 30,000 lb_f of impact (Shaver Manufacturing, 2009). The aft of the PUP frame provides an alternative mounting platform rather than a customary Category I or II three-point hitch. At a weight of 385 lb_f, it is well within the 1,500–2,000 lb_f capacity of the vehicle platform (Purdue University, 2018). The well driver should be able to drive sand point wells sufficiently deep to pull water from Earth's natural aquifer storage of groundwater. The well-driving attachment will provide an increased utility to the PUP, providing people in developing areas with an alternative and safe solution to clean water.

2.3. MATERIALS

Apart from the driver, sand point wells require only a handful of materials to construct a complete well. Materials that are needed per well include one well point, well point couplings, one well point drive cap, and lengths of galvanized steel pipe. The heart of a sand point well is the well point. The well point is a 36 *in* galvanized steel spike, followed by a mesh for water withdrawal and a threaded portion to accept more plumbing. As illustrated in Figure 1.2, the point enables groundwater to be withdrawn through the pores of the mesh, preventing small gravel and sands through access. Well point couplings connect the point to the pipes and then couple other pipes together. Standard couplings cannot be used. Unlike standard pipe couplings, well point couplings allow for connected pipes to be butted end to end. Standard couplings would carry the impact load through the threads only, easily breaking or damaging them. Galvanized steel pipe will connect with the couplings to form the required well depth with the drive cap at the top. The drive cap is responsible for preventing thread damage to the uppermost pipe while driving. For two-inch pipe, the cap is approximately four inches in diameter, spreading out the load and allowing a greater impact contact area. The cap will need to be swapped with the next section of pipe as the well is driven into the ground. The cap may be reused for many wells assuming it is in usable condition, as it is not a consumable item.

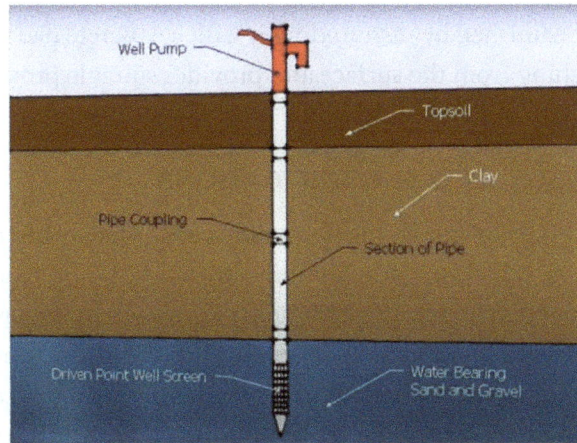

FIGURE 1.2. A closer look at a sand point well (SimpleSolar Homes, 2013)

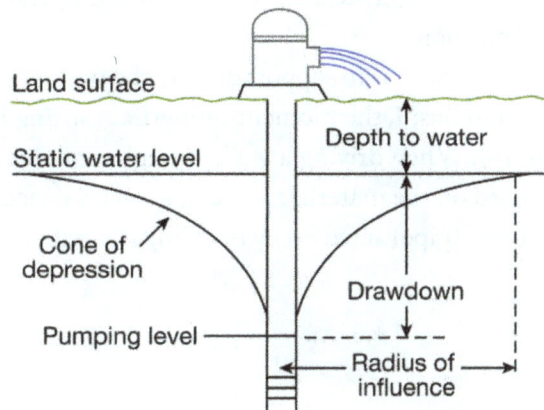

FIGURE 1.3. Water drawdown effect (Kansas Geological Survey, 2001)

2.4. DRIVING PROCESS

The operator must first choose a suitable well location, ideally in a sandy soil and away from sources of surface contamination. Sandy soils are less likely to contain large underground boulders that might break a point. Next, the operator must move the vehicle approximately in position. The post driver can be raised from the travel position to the operating position and then microadjusting vehicle position as needed. Next, the wheels must be chocked and the support arms of the post driver lowered. The operator must initially set a well point coupled to a five-foot length of galvanized pipe and a drive cap moved to the upper section. After each progressive drive, new sections of pipe must be coupled to driven pipe, and the cap must be replaced. Detection of water in the driven well can be done quickly by a weighted string and measuring the depth of the water table. Ideally, the completed well should be several feet below the water table to allow for a "drawdown" effect, illustrated in Figure 1.3. Based off the depth of well, a suitable water pump must be chosen. In general, a pitcher-style pump is used for shallow wells (25 ft or less in depth), or a submersible pump for deep wells can be used. The driven well must be finished by surrounding it with a concrete pad. A properly sized pad reduces the effect of contaminants leaching from the surface and provides suitable protection to the well casing.

2.5. MODIFICATIONS

To drive a plumb well without pipe buckling, many modifications were made to the vehicle and the driver mechanism. Modifications to the vehicle include running a hydraulic pump from an auxiliary drive, welding mounting links, and installing a lift cylinder. The hydraulic pump used in the prototype is chain-driven off a jackshaft and runs continuously when *rpm* climbs above engine idle due to the centrifugal clutch. The pump can produce pressures greater than 2,000 *psi*, easily supporting the highest demanded system pressure of 1500 *psi*.

Mounting links were welded to the rear of the vehicle frame to support the post driver. The links enabled operation in the vertical orientation but also allowed for the driver to be lowered to its road travel orientation (about 20° from horizontal). A two-inch bore hydraulic lift cylinder raises and lowers the driver in position. The lift cylinder is incorporated easily into the existing hydraulic circuit. The ability to move into a road travel mode enables better vehicle balance in travel, critical for the safety of the operator. Figure 1.4 shows a mounted post driver oriented in the vertical position.

The post driver needed several modifications to support driving the two-inch pipe without buckling. Buckling is affected by E, the Young's Modulus; I, the moment of inertia resisting buckling; and L, the length of the slender member (Matthew, 2016). When driving a well, Young's modulus and the moment of inertia of the driven section are constants based on the materials as well as its cross-sectional shape. Therefore, changing the effective length of the member (pipe) acted on by buckling is a critical variable.

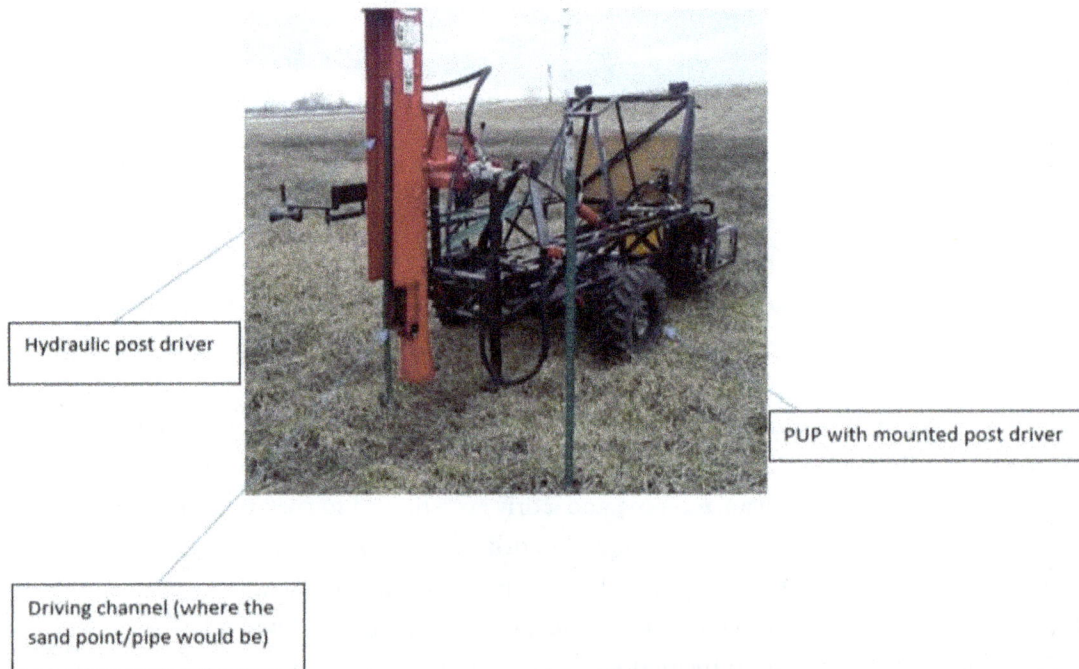

Hydraulic post driver

PUP with mounted post driver

Driving channel (where the sand point/pipe would be)

FIGURE 1.4. An unmodified post driver attached to a 201x era PUP

FIGURE 1.5. Shaver's steel post sleeve versus well-point drive sleeve

Lengths of pipe up to 80 *in* will be subjected to large impacts, bending an inadequately supported well casing. A drive sleeve comparable to Shaver's HD-8 steel post setting sleeve was scaled up and fabricated to fit the well-driving apparatus. Pieces of this new device and the Shaver's unit are shown in Figure 1.5. The drive sleeve supports the top of the driven well, keeping it centered in the drive channel and preventing the drive point cap from wandering during impact. Midway down, approximately 30–40 *in* from the ground, another support has been added. Supporting the pipe toward the middle will be crucial to minimizing the effective member length, therefore increasing the critical buckling load. This midsupport will cancel any side loading forces but will allow free movement in the vertical axis.

2.6. TESTING

Various tests before and after vehicle modification serve as critical guidelines to the success of the well-driver attachment. Checkpoints include driving a steel post, driving a larger wood post, and finally driving three wells in local areas. Driving a steel post successfully, without bending, will be a good first test for the driver. Steel posts will be more likely to bend than steel pipe due to their decreased moment of inertia. The steel posts have a small surface area to drive into the ground, confirming the operation of the driver.

Driving wood posts will be difficult but in a different aspect than steel posts. Wood posts (4″x6″ timber for test) have a large surface area to drive into the soil, and the rough sides provide greater resistive friction during driving. This test will ultimately confirm the power of the driver. Assuming it has the power to drive a 4″x6″ (driven SA: 19.25 *in²*) post, it should be able to drive two *in* galvanized pipe with no problems (driven SA: 3.14 *in²*).

Finally, driving the wells will be testing the practicality of meshing a driven well with the post driver and analyzing the water quality that can be obtained with this method. Driving the well point and the galvanized pipe will test the modifications to the driver as well as integration for the various modifications. Water quality will be analyzed by a third-party professional well drilling company for each of the three water well locations.

Samples will be compared to US water quality standards, checking pH, total dissolved solids, hardness, iron, iron bacteria, manganese, copper, nitrates, sulfides, and tannins (AquaScience, 2017).

2.7. LIMITATIONS

Installing sand point wells is not logical in all areas; there are limitations to these wells in placement and pump performance. Due to the nature of driving a well, the placement of the well is critical. The well-driver operator must choose a sandier soil, preferably one known with a higher water table and far away from potential sources of contamination. Sandier soils will give the best water-drawing performance from the well point and reduce the likelihood of hitting an underground boulder. Pump performance should also be considered. The well owner may use a pitcher-style pump for wells 25 *ft* or less in depth and an electric narrow-diameter submersible-type pump for depths up to 100 *ft*. A user concerned about pump output or cost may wish to further investigate water table depths and alternative pumping options.

3. IMPACT

3.1. HEALTH

Successfully implementing a powered well-driving rig into developing areas adds a level of health and water security. Raising the quality and quantity of water available will create big positive changes locally. Users will be less likely to contract waterborne viruses and will rise to new standards of living. The young and the elderly will have reduced risk of illness, contributing to a higher infant survival rate and a longer average lifespan. The increase of water available per capita enables more diverse uses of water, as compared to only fulfilling basic needs. Better cleaning and hygiene and a better overall state of well-being immediately improve the area's quality of life.

3.2. SOCIAL

One of the largest impacts of having a clean, reliable water source will be its social benefits. The daily time spent retrieving water could be used for activities such as going to school and earning greater income. These activities, once limited by retrieving water, will contribute to incremental benefits on local and national levels. Current citizens will have more time available for productive activities, and future generations will turn out a more skilled workforce and improve the national economies of the developing world as they become increasingly educated. Improving the source of water brings a new outlook on life and opens a world of possibility for millions living today in a water-scarce environment.

3.3. SMALL BUSINESS POTENTIAL

An entrepreneur may be able to turn a well-driving rig into a profitable business model. The modified post driver coupled with an appropriate power source could bring revenue to an enterprising individual by renting or operating the well-driving rig. A market exists for not only communities also farmers. Driven wells have the possibility of providing irrigation where it had previously been impractical and improving crop yields. The marketplace in this sector is large and diverse. The business opportunity itself presents greater interest, as the entrepreneur may also serve as an importer of driven well parts, effectively cutting out the middleman. The entire operation may pay for itself in a few years or even a few months' time. There is a great urgent need for these types of entrepreneurs and the services they could provide on a community-by-community level within the developing world.

4. CONCLUSION

Water sourcing for developing areas is a complex problem and requires a set of unique solutions to better meet the needs of users. Until a proper infrastructure is developed, water supply will require small-scale water works projects on a community-by-community basis. The well-driver PUP fills an exclusive market niche, being a sustainable, affordable well installation method. Operators of the system don't require complex training or need a formal education. The powered well-driving process is a solution for the people and by the people of developing areas.

5. WELL DRIVER PUP QUESTIONS

1. Describe and characterize the need for freshwater in the communities of the developing world, along with the social constraints currently limiting the expansion of adequate freshwater systems for communities.
2. Describe a shallow tube (sand) well, including its components and installation process.
3. Describe the mechanical constraints present in the driving process that require stabilization for effective installation of the pipe into the soil.
4. Identify the current limitations to tube well installations and describe how the well-driver PUP systems overcome some of these issues. Which limitations remain unaddressed?
5. Describe the advantages of using a well-driver PUP system for the installation of tube wells compared to traditional drilled water well installations.

6. REFERENCES

AquaScience. (2017). *Request a free well water test kit.* Retrieved May 15, 2018, from https://www.aquascience.net /free-water-test/.

Environmental Literacy Council. (2015). *Water in developing countries.* Retrieved May 17, 2018, from https://enviroliteracy .org/water/water-supply/water-in-developingcountries/.

Kansas Geological Survey. (2001). *KGS—Reno County geohydrology—Ground water recovery.* (2001). Retrieved May 17, 2018, from http://www.kgs.ku.edu/General/Geology/Reno/gw03.html.

Matthew, T. (2016). *Engineering core courses C5.1 Euler's buckling formula.* Retrieved May 12, 2018, from http://www .engineeringcorecourses.com/solidmechanics2/C5-buckling/C5.1-eulers-buckling-formula/theory/.

Purdue University. (2018). *Purdue Utility Project (PUP).* Retrieved May 10, 2018, from https://engineering.purdue.edu/pup/.

Shaver Manufacturing (2009). *operator's manual: Hydraulic post driver model HD-8 & HD-8-H.* Retrieved April 19, 2018, from http://shavermfg.com/media/uploads/HD8-Operator-Manual.pdf.

SimpleSolarHomes. (2013, March 20). *How to install a driven sand point well.* Retrieved March 20, 2018 from https:// www.youtube.com/watch?v=I-9g6iZGkoY.

Zackariah Horn, Dr. Robert M. Stwalley III

The Problem & Current Solutions

[1]

- 1 in 9 people world wide do not have access to safe and clean drinking water [2]
- Water hauling tasks are primarily given to women, and surface waters are often contaminated
- Current water access practices include chlorination, hand-dug wells, or professional wells
- A mechanical solution to access groundwater is proposed

Objectives

- Prototype an attachment for a PUP (Purdue Utility Platform) for groundwater access
- Remain cost effective
- Design with little to no experience for safe operation

PUP Attachment

- Modified post driver
- Provides up to 30,000 lbs. of impact force (at full stroke)
- Attaches easily to vehicle
- Retains compatibility with PUP accessory drives and weight capacity
- Cost effective, low maintenance design

- Removes strenuous and dangerous conditions for laborers
- Requires only basic mechanical knowledge to operate and troubleshoot
- Quick installation time (hours vs. days/weeks)
- Wells can be installed in difficult to access locations (as compared to larger drilling rigs)

Sand Point Wells

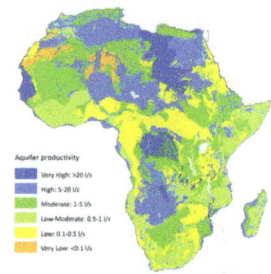

[3]

- Sand point wells draw from Earth's natural groundwater storage
- Reasonably protected from leaching surface contaminants
- Requires few specialty components

Impact

[4]

- Sustainable water access solution
- Increased time for education
- Reduced infant/ newborn mortality rate
- Potential small business endeavor, regional economic boost

Images/References:
[1] Wappelmann, A. (2016, April 13). Has Haiti's Cholera Epidemic Become a Permanent Problem? Retrieved April 2018, from https://news.ufl.edu/articles/2016/04/has-haitis-cholera-epidemic-become-a-permanent-problem.php
[2] The Water Project. (2016, August 31). Facts and Statistics about Water and its Effects. Retrieved April, 2018, from https://thewaterproject.org/water-scarcity/water_stats
[3] British Geological Survey. (n.d.). Download Digital Groundwater Maps of Africa. Retrieved June 19, 2018, from http://www.bgs.ac.uk/research/groundwater/international/africanGroundwater/mapsDownload.html
[4] African Children's Choir. (2018). Give Now – African Children's Choir Retrieved May 30, 2018, from https://africanchildrenschoir.com/help/give-now/

Acknowledgements:
Purdue ABE Department
Dr. John Lumkes
Scott Brand

PURDUE AGRICULTURE

PURDUE ENGINEERING *Think Impact.*

Purdue University is an equal opportunity/equal access institution.

CHAPTER 2. IMPROVEMENTS TO WELL-DRIVER PUP

This article continues to detail the evolution of the well-driver Purdue Utility Platform (PUP) vehicle and its initial round of water well tests. Initial fence post–driving tests concluded that the vehicle's suspension suppressed the driver's impact, and four hydraulic stilt outriggers were added to each corner of the vehicle to lift and level the unit, with the tires off the ground. Following this modification, both steel and wooden fence posts were successfully driven. Other minor modifications to the device were completed following the fence post tests, and water well installation under the rules established by the State of Indiana commenced. Five test water wells were driven in Montgomery County, Indiana. Although water was found in several of the wells, only one was a truly successful installation. A 7 *m* deep well on sandy soil near New Richmond was developed and finished as an active well. It had a capacity of about 11 *L/min*. After initial disinfection, well water–quality checks conducted by the Montgomery County Health Department indicated that the well was sanitary and safe for human use. This paper was originally presented at the 2019 annual international meeting of the American Society of Agricultural and Biological Engineers. The poster presented at the meeting is also included.

A LOW-COST MECHANIZED TUBE WELL INSTALLER

ZACKARIAH J. HORN (PURDUE UNIVERSITY AGRICULTURAL &
BIOLOGICAL ENGINEERING) AND ROBERT M. STWALLEY III

ABSTRACT

People in developing countries are often limited in the amount of safe, clean water they can provide for their families. Water sources may be far away and/or contaminated, while limited finances can make traditional methods of water access infeasible. To combat these water access problems, a mechanized means to install sustainable driven water wells was developed. A modified hydraulic post driver was mounted to a Purdue Utility Platform (PUP) vehicle to provide the impacts necessary for driven water well installation. PUP vehicles, which can be thought of as a hybrid between a light truck and a tractor, are currently performing utilitarian missions in developing countries as they pull tillage implements and power various attachments. PUPs are built using commonly available local materials in a region and are constructed with minimal tooling. The PUP provided hydraulic power to the post driver to drive water wells, enabling it to replace the impacts commonly delivered by hand tools. Five wells were installed at sites across West-Central Indiana, with the deepest well being driven 23 ft. This suggests that the well-driver PUP could install shallow driven wells, and the connection with the PUP could make this a viable water access solution in developing countries. These low-cost wells could provide safe, clean water on a communal basis, increasing the level of hygiene and reducing pathogens that might be present in unimproved water access methods. Design and testing of the well-driver PUP prototype will be discussed, and the details regarding the successful well will be provided.

KEYWORDS: developing countries, driven well, post driver, PUP, Purdue Utility Project, sustainable, water, water access, water well, well driver

1. INTRODUCTION

1.1. THE PROBLEM

Developing countries often have problems that are associated with water quality and/or water quantity. Lack of funds and skilled laborers leave developing areas with no infrastructure or sustainable method of safe, direct water access. Families often carry water from great distances to fulfill their basic drinking and cooking needs. Surface waters and shallow groundwater may often be contaminated by the common practices of open dumping waste or from agricultural runoffs. These contaminants may pose great health concerns to those who ingest them. Water-borne diseases account for as much as 80% of illness and death in the developing world (United Nations, 2003).

1.2 CURRENT WATER ACCESS SOLUTIONS

Several types of water access are commonly found in developing countries, ranging from archaic methods to the most technologically advanced. One of the most primitive forms of water access is surface waters. Water may be collected from sources such as lakes, rivers, and streams. Surface waters may be prone to contaminant entry by either surface runoff or shallow groundwater leaching.

Hand-dug water wells are very labor intensive, requiring workers to dig their way past the groundwater table. Oftentimes, these constructions are done in the summer months when the water table is at its lowest, ensuring a continuous supply of water year-round. Hand-dug wells may be susceptible to contaminants entering via runoff into the top of the well or even through leaching through an imperfect permeable well casing if one happens to be installed.

Hand auger wells are essentially a manual way of "drilling" a water well. An auger is connected to lengths of drive stem pipes, with workers providing the torque for drilling above ground. This method of water access may be depth limited, especially if rocky or clay-like soils are encountered. It may be prone to contamination by shallow groundwater leaching if not equipped with a plastic or galvanized steel well casing to an acceptable depth.

Driven wells are composed of a well point and lengths of galvanized steel pipe to serve as a well casing. A laborer with a sledgehammer hits the top of the driven well through a well point drive cap. A driven well can be very labor-intensive to install and faces the same depth limitations as the hand auger wells. Driven wells may also be difficult to install in tough soils, and they can be more prone to contaminants if the well is too shallow.

Drilled water wells are the most ideal form of water access, but they come with a price. The profound cost of a drilled well makes this method the least attractive and the most difficult to justify in a development setting. Drilled wells require specialized equipment and massive drill rigs mounted to commercial-sized trucks, in addition to trained operational crews.

1.3. PROPOSED SOLUTION

To address the problems with water quality and access, a powered way to install driven water wells was devised using a repurposed fence post driver. The solution needed to be machine powered, remain low cost, require no formal education, and provide a long-lasting sustainable impact in the area of implementation. It was hypothesized that incorporating machine power into the well-driving process would likely place the well point into deeper water-bearing soil formations, thus reducing the possibility or amounts of contaminants in the tapped water supply. A Shaver HD-8® hydraulic post driver was mounted to the rear of a PUP frame to deliver hammering impacts for the driven water wells.

2. WELL-DRIVER PUP

2.1. PUP AND ATTACHMENTS

PUPs are a low-cost utility vehicle designed for developing areas and are built using commonly available parts and materials to the region. PUPs can be thought of as a smaller-sized hybrid between a truck and a tractor. PUP vehicles serve utilitarian purposes such as running water pumps, maize grinders, and tillage implements as well as the ability to haul as much as 2,000 lb_f of payload (Purdue University, 2019). These machines are typically built from stock steel shapes and contain used but usable automotive parts. The proposed PUP attachment fits within the spirit of the utility purposes of the platform and provides users with an engine-powered means to install driven water wells at a location of their convenience or need.

2.2. WELL-DRIVER ATTACHMENT

The well-driver attachment and components are powered with a hydraulic pump connected through the powertrain off a diesel engine. The hydraulics provide motive power to the post driver, a tilt cylinder, and four hydraulic outriggers mounted on each corner of the vehicle. The post driver provides a means of impact to the top of a driven well, replacing a manual laborer or laborers using a sledgehammer. The Shaver HD-8® post driver features a 53.5 *in* stroke impact ram, which is hydraulically lifted and gravity-plus spring-powered down. The driver provided an effective weight of 360 lb_f to the spring-powered driving ram (Shaver, 2009). The tilt cylinder enabled the post driver to transition from its transport position into operational position as the cylinder was extended. The tilt cylinder allowed the PUP to have greater stability off-road, primarily by lowering the vehicle's center of gravity and slightly improving the wheel weight distribution. Four hydraulic outriggers were welded to the PUP, one toward each "corner" of the vehicle. The outriggers eliminated any suspension play from the vehicle during driving operations, and they removed any tendency for the driver to deliver unsquare hits to the driven well pipe cap.

2.3. WELL-DRIVING ACCESSORIES

To deliver proper impacts to a driven well assembly, the post driver had to be outfitted with various accessories. First, a well cap drive sleeve was fabricated, which was essentially a large version of the Shaver steel post drive sleeve. The well cap drive sleeve was built using 4½″ outer diameter round tubing, a ½″ plate steel, and ½″ steel rods. The tubing provided lateral support to the drive cap with about ¼″ of clearance, and the plate steel provided placement for a good strike. The steel rods offset the drive sleeve from the H beam on the driver, allowing the center of the sleeve to remain centered relative to the post driver's strike plate.

Safety chains were added to aid drive sleeve retention, preventing the sleeve from leaving the inside of the drive channel on the post driver. The ¼″ link chain was welded to one side of the driver, and steel hooks were welded on the opposed side as a quick method of securing the drive sleeve. Four safety chains were evenly spaced along the stroke of the post driver to ensure operator safety during each stage of a well drive. Figure 2.1 shows the well-driver PUP in operational position.

3. TESTING AND RESULTS

3.1. PRELIMINARY

3.1.1. STEEL POSTS

Common steel fence posts were first driven as a maiden test for the hydraulic post driver. Steel posts would show if there were any potential problems with buckling, as the posts are thin in cross-section and have a large height to width ratio. With the addition of the hydraulic outriggers, steel fence posts were driven squarely and did not buckle under impact loading from the driver. Multiple posts were tested, and they were easy to drive straight and plumb.

FIGURE 2.1. Well-driver PUP in operational position

3.1.2. WOOD POSTS

Several 4″x6″ wood posts were driven to determine the impact power available from the driver and whether the PUP could withstand the loadings delivered to its chassis. The Shaver HD-8® post driver was designed to mount to either a skid steer or a tractor; thus, the well-driver PUP was a fairly light mobile machine, at just over 2,400 lb_f wet weight. The post driver provided enough power to drive the 4″x6″ wood posts (bottom ends angle-cut to 45°) more than three feet into the ground, with no noticeable "hop" from the vehicle.

3.2. DRIVING WATER WELLS

3.2.1. LEGALITY

Prior to driving water wells in Indiana, the legality was examined. All of the prospective sites in which landowner access was granted were in Montgomery County, Indiana. According to the Indiana Department of Natural Resources, if a driven well is 2″ or greater in diameter or more than 24 ft deep, a licensed well driller has to install the well. As a 2″ diameter material was chosen, the principal investigator received his Indiana well driller's license to be in accordance with the law. No well permit was needed in Montgomery County, which sped up the process and testing. Finally, the Department of Natural Resources required water wells to be registered within 30 days of the well completion.

3.2.2. SITES 1-4

Water wells were installed in various rural locations in the Montgomery County, Indiana, area. Sites of interest were chosen based on a few factors such as landowner permission, likelihood of finding water, and site accessibility by transport, a full-sized pickup truck and 16″ trailer.

Site 1 was located in Darlington, Indiana. The area was thought to have a high water table as when it was historically farmland, 40–50 years ago. The well-driver PUP installed a driven well (well 1a) to a depth of 13 ft and had a three ft static water level in the bottom of the well. The Waterra Super Twister submersible pump was lowered into the well casing, but it could not reach the static water level. The well point or lower section of pipe likely deflected and bent due to hitting an obstacle, possibly a boulder.

In an effort to pump water from Site 1, a second well was installed approximately 200 ft from the first well. A second water well at Site 1 (well 1b) reached a depth of 10 ft, and one ft of static water remained in the bottom of the well. Water was able to be pumped intermittently but not continuously. The test pump moved 3.5 gpm at 10 ft, which emptied the static water level in the well in approximately three to four s. The well needed to recharge before any additional water could be pumped.

Site 2 was located in New Richmond, Indiana, near Coal Creek. The soil appeared to be somewhat sandy. Well 2 was installed to a depth of 23 ft with a static water column of 20 ft in the well casing. The submersible pump lowered into the well casing provided a continuous supply of water at approximately 3.0 gpm.

A third well location, Site 3, was located in Linden, Indiana. A well was driven 17 ft total, with no discernable static water level. After driving through a confining layer at approximately 9–11 ft (likely clay), another confining layer was hit at approximately 15 ft. The driver reached its capabilities at 17 ft, when impacts failed to push the well point any deeper in the soil. As the point was buried in an impermeable soil medium, water was unable to be pumped from this location.

Site 4 was the final driven well installation. This site was located in New Richmond, Indiana. A well was driven to 23 ft deep, but the well point was driven beyond the water-bearing soil formation. One ft of static water appeared to be in the bottom of the well at all depths. Therefore, it was unknown precisely at what depth groundwater was hit, but the layer was clearly thin, as the well point quickly passed through it.

4. RESULTS

At every site, a water-bearing soil formation was struck. A summary of the driven well statistics can be seen in Table 2.1. Although water was not able to be pumped continuously at all sites, the presence of water in the

TABLE 2.1. *Installed well details*

WELL #	WELL DEPTH *(FT)*	STATIC WELL COLUMN *(FT)*	CONFINING LAYERS *(FT)*
1a	13	3	N/A
1b	10	1	N/A
2	23	20	N/A
3	17	N/A	9–11, 15–17
4	23	N/A	N/A

FIGURE 2.2. Pumping clear water from Test Well 2

well casing of all driven wells displayed proof of concept. As Forrest Wright stated in *Rural Water Supply and Sanitation*, "There is no infallible method of locating a successful well before drilling. All methods fail at times to produce satisfactory results" (Wright, 1977). This suggests that groundwater access is a complex endeavor and relies on a multitude of variables. Wells 1a, 1b, 3, and 4 were filled and abandoned properly.

4.1. WELL DEVELOPMENT

Driven water wells must be properly developed to produce safe, high-quality water. First, the well must be surged. As Waterra states, "Surging is the process of creating a larger than usual flow through a screened interval usually with the goal of cleaning out smaller particulate matter or biological matter from the filter pack and/or adjacent formation" (Waterra, 2018). Well surging was accomplished at Site 2 by an inertial pump from Waterra. Additionally, surging helped to reduce the turbidity, or presence of suspended particles. Figure 2.2 shows clear water flowing from Well 2.

Once the well was deemed to be adequately surged, it had to be disinfected. Disinfecting a water well can be done by shock chlorination, a process in which chlorine is added in the well casing to reach a concentration of 100 *ppm* (parts per million). The concentration of chlorine, typically achieved through the addition of household bleach, kills bacteria colonies in the water.

4.2. WATER TESTS

Water tests were conducted after well disinfection by the Montgomery County Health Department in Crawfordsville, Indiana. P/A (presence/absence) tests were conducted for total coliform and *E.coli*. Total coliform serves as an indicator organism. Its presence, especially in large amounts, generally indicates an unsafe water specimen.

Three water tests were done at test Well 2. The first test was labeled satisfactory and bacteriologically safe. A second well test was submitted a few weeks later but was ruled unsatisfactory. It was speculated that the re-introduction of the pump (likely no longer disinfected) or a regrowth of a bacteria colony was likely the root cause of the contamination. The well was redisinfected, and the final (third) water test was satisfactory and bacteriologically safe. The American Groundwater Trust suggests that multiple shock chlorination attempts may be necessary, depending on the strength of bacteria colonies in the water well (American Ground Water Trust, 2012). Further monitoring of the quality of this well will be performed in the future.

5. POTENTIAL IMPACT

The impact from the well-driving PUP attachment could reach a multitude of sectors in the countries of the developing world due to the need for access to safe, clean water with shorter distances for carrying water. Installing low-cost driven-type water wells could provide health benefits, education benefits, and economic benefits in the region of implementation. Assuming a properly developed and disinfected well, there should be reduced waterborne pathogens in the drawn water. Reduced pathogens decrease water-related diseases. The time needed to collect water should be able to be decreased, lessening the tendency of a joint or skeletal injury by those who carry water across great distances.

Education benefits resulting from safe and close water access points are greatest for children, especially young girls. Children are commonly assigned the task of fetching water, which can often be a time conflict while receiving an education. The World Health Organization states that "girls often miss out on an education because they have to help with household chores and, when money is scarce, it's usually the boys who get chosen to go to school" (World Health Organization, 2001).

Economic benefits of increasing water quality and access are generally a result of improved health and education. Overall, improved health of a region's people enables workers to be more productive and reduce the number of sick days taken by workers. Studies have shown that increasing education has a correlation to increased GDP of a country (Hanushek & Wößmann, 2007). On a regional scale, healthier and more educated workers generally have propitious effects to their economy. Within a developing region, the well-driver PUP could be a significant instrument to improve the quality of life.

6. CONCLUSION

The well-driver PUP attachment has been shown to successfully install driven water wells in West Central Indiana. Due to the PUP's current utilitarian role in developing countries, the attachment may be able to be used in these regions. Furthermore, well-driver operators would only require minimal education, all of it relative to the machine operation and best practices. The well-driver attachment potentially brings a sustainable, low-cost solution for water access to developing countries around the world. Further work will examine the possible variation in well sizes, the continuing quality of the wells, and design optimization of the well-driver attachment.

7. WELL-DRIVER PUP IMPROVEMENT QUESTIONS

1. Explain why the vehicle suspension mechanism on the PUP dampened the driving impact.
2. Describe how a four-cornered leveling system improved the installation process and shortened the on-site preliminary setup time necessary to begin operations.
3. Why was certification as an Indiana-approved water well installer necessary before actual water wells tests could be initiated?
4. Describe in detail the process of developing a well following the installation of the casing into the ground. Why is this process important to the long-term productivity of the well?
5. Explain why although three of the first five initial trial water wells contained water, only one was considered successful and worthy of development.

8. REFERENCES

American Ground Water Trust. (2012). *Water well disinfection procedure.* https://agwt.org/content/water-well-disinfection-procedure.

Hanushek, E. A., & Wößmann, L. (2007). *Education quality and economic growth.* https://siteresources.worldbank.org/EDUCATION/Resources/278200-1099079877269/547664-1099079934475/Edu_Quality_Economic_Growth.pdf.

Purdue Engineering. (2019). *Purdue utility project (PUP).* https://engineering.purdue.edu/pup/.

Shaver. (2009). *Operator's Manual: Hydraulic post driver model HD-8 & HD-8-H.* http://www.shavermfg.com/media/uploads/HD8-Operator-Manual.pdf.

United Nations. (2003). *"Water-related diseases responsible for 80 per cent of all illnesses, deaths in developing world," says secretary-general in Environment Day message.* https://www.un.org/press/en/2003/sgsm8707.doc.htm.

Waterra. (2018). *Application: Surging.* http://www.waterra.com/pages/applications/BodyMovin/surging.html.

World Health Organization. (2001). *WHO World Water Day report.* https://www.who.int/water_sanitation_health/taking charge.html.

Wright, F. B. (1977). *Rural water supply and sanitation* (3rd ed.). Robert E. Krieger Publishing.

PROTOTYPING A WELL-DRIVER PUP (PURDUE UTILITY PROJECT) TO INSTALL DRIVEN WATER WELLS

PURDUE UNIVERSITY

Agricultural & Biological ENGINEERING

Zackariah Horn & Dr. Robert M. Stwalley III

The Problem & Current Solutions

[1]

- 1 in 9 people world wide do not have access to safe and clean drinking water [2]
- Water hauling tasks are primarily given to women
- Surface waters are often contaminated
- Current water access practices include chlorination, hand-dug wells, or drilled wells
- A mechanical solution to access groundwater is proposed

Objectives

[3]

[4]

- Provide a powered method to install driven water wells
- Require minimal training
- Remain low-cost
- Provide a sustainable, lasting impact

Driven Water Wells

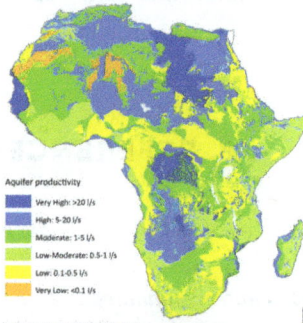

Aquifer productivity
- Very High: >20 l/s
- High: 5-20 l/s
- Moderate: 1-5 l/s
- Low-Moderate: 0.5-1 l/s
- Low: 0.1-0.5 l/s
- Very Low: <0.1 l/s

[5]

- Driven wells draw from Earth's natural groundwater storage
- Wells of proper depth are reasonably protected from contaminant leaching
- Requires few specialty components

PUP Attachment

[6]

- Modified Shaver HD-8® hydraulic post driver
- Drives wells up to 25 feet
- Quick installation
- Adds utility and capability to PUP
- Cost effective and low maintenance

Testing & Results

- Installed wells in West Central Indiana
- Drove wells up to 23' deep

[7]

Impact

[8]

- Sustainable water access solution
- Increased opportunity for education
- Reduction of waterborne pathogens
- Potential small business

Images/ References:

[1] Weppelmann, A. (2016, April 13). Has Haiti's Cholera Epidemic Become a Permanent Problem? Retrieved from https://news.ufl.edu/articles/2016/04/has-haitis-cholera-epidemic-become-a-permanent-problem.php
[2] The Water Project. (2016, August 31). Facts and Statistics about Water and Its Effects. Retrieved from https://thewaterproject.org/water-scarcity/water_stats
[3] Purdue Engineering. (2019). Purdue Utility Project (PUP). Retrieved from https://engineering.purdue.edu/pup/
[4] Wright, F. B. (1977). Rural Water Supply and Sanitation (3rd ed.). Huntington, NY: Robert E. Krieger Publishing Company.
[5] British Geological Survey (n.d.). Download Digital Groundwater Maps of Africa. Retrieved from http://www.bgs.ac.uk/research/groundwater/international/africanGroundwater/mapsDownload.html
[6] Shaver. (n.d.). Standard Post Drivers. Retrieved from http://www.shavermfg.com/standard-post-drivers
[7] Driving a Well with a Well-Driver PUP [Personal photograph taken in Darlington, IN]. (2019).
[8] African Children's Choir. (2018). Give Now – African Children's Choir. Retrieved from https://africanchildrenschoir.com/help/give-now/

Acknowledgements:
Scott Brand
Dr. John Lumkes
David Wilson
Purdue ABE

PURDUE AGRICULTURE

PURDUE ENGINEERING Think Impact

Purdue University is an equal opportunity/equal access institution.

CHAPTER 3. ENHANCEMENTS TO WELL-DRIVER PUP CAPABILITIES

The post driver selected for use with the Purdue Utility Platform (PUP) vehicle was the smallest unit available from the manufacturer. This decision was based not on price but on the relative weight of the unit in comparison to the PUP. This meant that the driving impact force of the unit was originally limited to a combined gravitational and spring total of 1,600 N. This theoretically constrains the maximum depth of the device in well driving to about 15 m. While larger post drivers should be able to achieve greater depths, the additional weight of these units climbs much faster than their potential impact force. To overcome this limitation, the Purdue team sought to add additional weight directly to the driving ram, leaving the remainder of the driver assembly unchanged. Brackets and locking assemblies capable of holding an extra 440 N per side were added to the outside of the driving ram. This paper discusses the design and numerical analysis of the added weight brackets and describes a proposed testing program for the experimental verification of the potential depth enhancement. This paper was originally presented at the 2022 annual international meeting of the American Society of Agricultural and Biological Engineers. The slides presented at the meeting regarding this program are also included at the end of the chapter.

IMPROVING THE DRIVING CAPABILITIES OF A WELL-DRIVER PUP (PURDUE UTILITY PROJECT) TO INSTALL LOW-COST DRIVEN WATER WELLS

GRACE L. BALDWIN KAN-UGE, TYLER J. MCPHERON (PURDUE UNIVERSITY AGRICULTURAL & BIOLOGICAL ENGINEERING) AND ROBERT M. STWALLEY III

ABSTRACT

Globally, 1.8 billion people use an unimproved source of drinking water with no protection against contamination from feces. Improved sanitation from safe water and good hygiene could prevent around 842,000 deaths each year. This work addresses the sixth United Nations Sustainable Development Goal (SDG), which will help drive progress across many other SDGs:,ensuring the availability and sustainable management of water and sanitation for all, also known as WASH (water access, sanitation, and hygiene). WASH programming specifically creates access to safe water for drinking that is free from chemical and biological pollutants; sanitation, which includes access to a toilet (latrine) that safely separates human excreta from humans; and hygiene, or a focus on public health and the transmission of fecal-oral diseases. In developing countries, water access is not always available. This research seeks to improve access to the water component of WASH programming. In many locations around the world, people use unsafe water sources and lack sufficient access to water for both drinking and domestic purposes. Particularly in sub-Saharan Africa, women and children walk great distances to obtain access to water. People must have equitable and affordable access to safe and sufficient water that is palatable and in sufficient quantity for both drinking and domestic purposes before any other long-term economic development or social improvement can occur. This research seeks to increase access to subsurface water by improving the driving capabilities of the well-driver Purdue Utility Platform (PUP) vehicle. Worldwide, there are many locations where the water table depth is within the 10–20 m range. These locations include sub-Saharan Africa, the Caribbean, South America, northern India, Asia, and parts of the Asia Pacific Islands. These locations are places where the well-driver PUP could potentially be utilized if sufficient depth can be demonstrated on a repeatable basis. Increasing the vehicle's achievable depth capability will increase the number of locations throughout the world that the vehicle could be used to access groundwater for those with limited to no current water access. This unique vehicle can improve safe water access in developing countries for drinking and domestic purposes and play a key role in enhanced WASH programming. This paper will thoroughly discuss the engineering solutions to improve the driving depth of the vehicle.

KEYWORDS: Africa, groundwater, Purdue Utility Vehicle, PUP, tube well, international development, sub-Saharan, water, water resources, well driver, wells

1. INTRODUCTION

"A hydraulic post driver mated to a PUP has been designed to mechanize the process of installing driven water wells, also called the Well Driver PUP" (Horn, 2019)." The PUP is a three-wheeled low-cost utility vehicle designed for use in developing countries. These vehicles are typically built in-country with only locally sourced available materials and minimal tooling, predominantly for sub-Saharan African countries, those being Guinea, Nigeria, Cameroon, Nigeria, Uganda, and Kenya (Purdue Utility Project, 2021). The well-driver PUP is pictured in Figure 3.1, and Figure 3.2 portrays the examples of other PUP vehicles. These vehicles are used primarily for transportation purposes, such as a truck, taxi, or miscellaneous hauling, in areas where normal vehicles are not appropriate due to the severe terrain and lack of road accessibility. These vehicles have been used as ambulances to transport individuals in need of medical services from extremely rural areas to hospitals and clinics. The vehicles have been used for transporting goods to and from markets and as school bus alternatives, garbage collection vehicles, and taxis. Tillage attachments, seeders, harvest heads, water pumps, generators, threshers, and maize grinders have been designed and tested to be powered by the PUP (Horn, 2019). Such implements have helped improve smallholder farmers' access to markets and improved the livelihoods of those in sub-Saharan Africa. The well-driver PUP could potentially reduce dependency on manual labor for driven well installation in developing countries and keep laborers safer (Horn, 2019). Development of the original unit was detailed by Horn and Stwalley (2018, 2019, 2020).

In locations where the number of drilling rigs is reduced and where labor is scarce, the well-driver PUP vehicle could be used as an instrument of economic development, providing microbusiness development opportunities focused on well installation. Installing low-cost driven wells would improve the availability, quality,

FIGURE 3.1. Well-driver PUP (Baldwin, 2022)

FIGURE 3.2. PUP vehicles (Horn, 2019)

and accessibility of water in locations lacking sufficient water supplies. In order to be successful, the well-driver PUP operation and components must remain low-cost, easy to maintain, and based on locally accessible materials as much as possible. It is intended that operation of the vehicle should require no formal training in well drilling and/or geology. In locations with an appropriate water table depth, this vehicle could decrease both the wait time to install and the final cost of installed wells (Horn, 2019).

2. LITERATURE REVIEW

2.1. POTENTIAL IMPACT OF THE WELL-DRIVER PUP IN THE DEVELOPING WORLD

Global estimates indicate that 30% of Earth's entire freshwater resources are found in the ground, while 68% is locked up in ice and glaciers (US Geological Service, 2016). Groundwater serves as the world's largest source of freshwater and plays a critical role in meeting the needs of people around the world (de Graaf et al., 2015). Groundwater is the primary source of drinking water and supplies water for agricultural and industrial activities worldwide (Wada et al., 2014). Groundwater is the portion of the total precipitation that soaks into the earth's crust (approximately one-third) and percolates downward into the porous spaces in the soil and rock, where it remains or finds its way out to the surface (Horn, 2019). Through testing, Horn (2019) demonstrated that the well-driver PUP could achieve depths of up to 7.0 m. Though not yet demonstrated through formal testing, Horn (2019) analytically determined that the vehicle should be capable of achieving depths of up to 15 m without significant changes. The depth to the water table varies throughout the world. Demonstrating the well-driver PUP's capability of achieving depths of up to 15 m would increase the number of locations throughout the world where the well-driver PUP could be utilized to access water for individuals who currently have little to no access.

A high-resolution global-scale groundwater model was developed by de Graaf et al. (2015). They created such a model, highlighting from their model simulations the current water table depth on a global scale. Figure 3.3 provides insight as to the water table depth below the land surface throughout the world. Worldwide, there are many locations where the water table depth is within 15 m, specifically in the 10–20 m range.

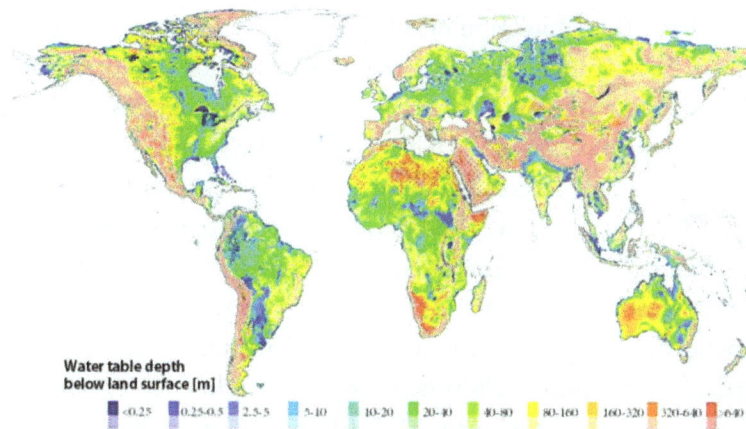

FIGURE 3.3. Water table depth below the land surface (de Graaf et al., 2015). The water table is most shallow in eastern North and South America; central Africa; and central Europe. It is deepest in western North and South America; Saharan and south Africa; southern Europe; the Arabian Peninsula; eastern Asia; and western Australia.

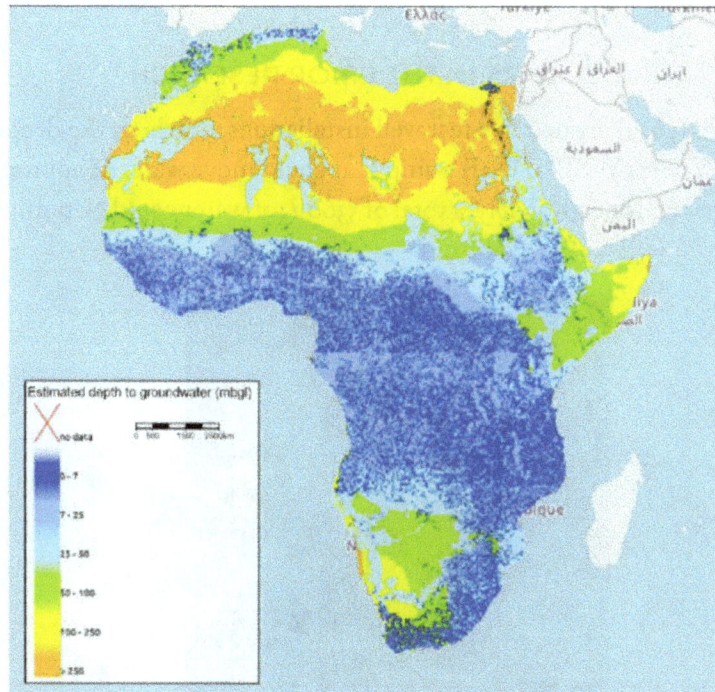

FIGURE 3.4. Estimated depth to groundwater (*mbgl*) and transboundary aquifer of Africa (International Groundwater Resources Assessment Centre, 2017). The groundwater is deepest in the Sahara and southwestern portions of the continent. It is shallower in sub-Saharan Africa.

Many of these locations are in sub-Saharan Africa, South America, northern India, Asia, and parts of the Asia Pacific Islands. These locations demonstrate potential places where the well-driver PUP could be utilized if sufficient depth can be demonstrated on a repeatable basis.

According to the United Nations, approximately 14.5% of the world's population is located within sub-Saharan Africa (United Nations, 2019). The world's population is projected to reach 9.8 billion people in 2050, and

11.2 billion by 2100 (Held, 2017). The population of sub-Saharan Africa alone is projected to double by 2050 (99%) (United Nations, 2019). Baldwin and Stwalley (2018, 2019) discussed the criticality of water supply and access to economic growth and community well-being. Of the 10 largest countries worldwide, Nigeria, located in sub-Saharan Africa, is growing the most rapidly (Held, 2017). Nigeria's population is projected to surpass that of the United States and become the third-largest country in the world shortly before 2050 (Held, 2017). Therefore, when considering locations where the well-driver PUP could impact the largest number of people, specific consideration should be given to locations in sub-Saharan Africa. The International Groundwater Resources Assessment Centre (IGRAC) and the UNESCO International Hydrological Programme have mapped the transboundary aquifers in Africa (International Groundwater Resources Assessment Centre, 2017). IGRAC, in collaboration with the British Geological Survey and the University College London, has developed maps to quantify the groundwater resources in Africa. Their results are highlighted zn Figure 3.4 and provide a high-resolution look at the depth to the water table in Africa. When specifically looking at sub-Saharan Africa, the approximate depth to the groundwater is predominantly within 25 *m* or less, followed by those in the 25–50 *m* range. This demonstrates the vast number of locations in sub-Saharan Africa where the well-driver PUP could access groundwater and has great potential to make a substantial positive impact for those with little to no water access.

2.2. GAPS IN THE PRIOR RESEARCH

The initial efforts of Horn (2019) created five test well installations, of which the deepest achieved a depth of 7 *m*. This well utilized a Waterra WSP-12V-3B pump that could move water at approximately 11.2 *L/min*. If a continuous duty cycle was run for eight *h/day*, a total of 5360 *L* of water could be pumped, providing water for

Ram Channel – (Grey), Plate Weld – (Yellow),
Weight Plate Weld – (Blue), Weight Support Plate – (Red)

FIGURE 3.5. Fusion 360 unloaded ram weight bracket design (Fidler et al., 2022)

FIGURE 3.6. Fusion 360 ram weight bracket loaded with weight plates (green) (Fidler et al., 2022)

a maximum of 268 people (Horn, 2019). Increasing the achievable depth of the well-driver PUP will expand the number of locations throughout the world where it could provide access to groundwater, as previously shown in Figures 3.3 and 3.4. To increase the driving capabilities of the well-driver, the weight of the driving ram should be increased. According to Driscoll (1986), the driver is likely able to achieve depths of up to 15 m, though this has not been confirmed through formalized testing. This depth is based on the effective weight of the spring-powered driving ram (Shaver, 2009). This indicates an achievable depth of 15 m in depth per the recommendation of Driscoll (1986) and Horn (2019).

3. METHODOLOGY

By increasing the ram weight of the driver, deeper depths can be achieved. The effective weight of the spring-powered driving ram is 1600 N (Shaver Manufacturing, 2009). The static weight of the ram was measured to be 1020 N. Therefore, the remaining 580 N is derived from the spring of the driving ram. The addition of brackets, welded to the lower portion of the driving ram, allows for individual weights to be added onto the driving ram to increase its overall weight. Each suitcase weight is 110 N and is made of A36 steel. A total potential of 440 additional N is possible per side, with a possible cumulative weight of 880 N. Since the static weight of the driving ram is 1,020 N, it is not advisable to test weight additions greater than that of the driving ram static weight. Testing up to a weight of 880 additional N allows for a factor of 1.15 to be maintained. To create the bracket design, 3D modeling, part drawings, and finite element analysis (FEA) were carried out in Fusion 360 (Fidler et al., 2022). The unloaded ram weight bracket design is displayed in Figure 3.5, and

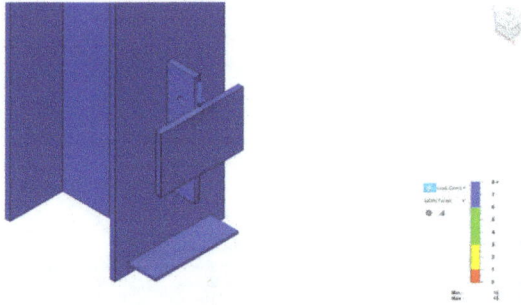

FIGURE 3.7. Safety factor Fusion 360 analysis (Fidler et al., 2022)

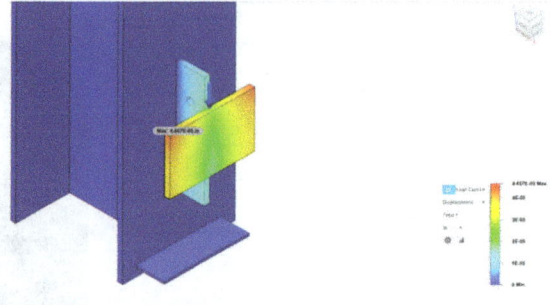

FIGURE 3.8. Deformation Fusion 360 analysis (Fidler et al., 2022) with maximum displacement of 4.657E-05 *in* or 1.2×10−4 *cm*

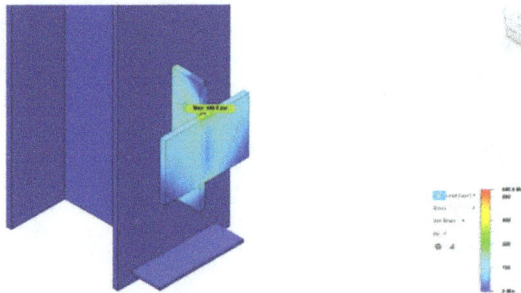

FIGURE 3.9. Von Mises Fusion 360 analysis (Fidler et al., 2022) with maximum stress of 650 *psi* or 4.5 *kPa*

FIGURE 3.10. Fabricated unloaded ram weight bracket design (Fidler et al., 2022)

FIGURE 3.11. Fabricated ram weight bracket design loaded with weight plate (Fidler et al., 2022)

the loaded bracket design is shown in Figure 3.6. The load case modeled consisted of a uniformly distributed dynamic load of 1,110 *N*, which was applied vertically to the top of the weight plate. From this analysis, the displacement, safety factor, and Von Mises stress reports were obtained.

4. RESULTS

The ram weight bracket design was intentionally overengineered to provide substantial support for its designed use (Fidler et al, 2022). The factor of safety for the design is 15, with an ultimate load of 16,690 N and a dynamic load of 1,110 N used. This high factor of safety allows for increased weight to be added if in the future higher weights are required for testing. The factor of safety outcome is portrayed in Figure 3.7. The deflection of the weight-supporting plate is near zero, with a maximum displacement of only 1.2×10^{-4} cm. These results are portrayed in Figure 3.8. The horizontal plate welded on the lower portion of the ram channel will help to partially support the weight plates. The results of the Von Mises stress analysis, presented in Figure 3.9, indicate that very minimal stress is likely to occur, with a maximum of only 4.5 kPa. This maximum stress will be even smaller once weld metal is applied, which will provide further stiffening to the joint (Fidler et al., 2022). The final design was fabricated, and the unloaded ram weight bracket design is shown in Figure 3.10. The left side of the figure shows the front portion of the drive channel, and the right side of the figure shows a side view of the final design. Figure 3.11 illustrates the driving ram raised, with the newly fabricated weight plates easily viewable. A side view of the driving ram is also shown with weight plates present. Following the successful fabrication and modeling of the weight bracket and weight plates, future field testing of the design is planned.

5. FUTURE TESTING

During the well-driving process, the well point remains in the ground once driven. Horn (2018) first demonstrated the driving capabilities of the well-driver through the installation of wooden fence posts. Alternatively, if wooden fence posts are used, these can be pulled back up using a tractor upon completion. Utilizing wooden fence posts allows for reduced experimental costs if a formal well point and piping section were used per treatment for the following experiment. Therefore, future testing will first focus on the installation of multiple wooden fence posts, with beveled ends to mimic a traditional well point. Five treatments will be used, with four wooden posts driven per treatment. The first treatment will consist of only the driving ram raw weight (1,020 N) and no additional weight added. This will allow for a benchmark measurement to be taken. For the second treatment, two weights will be added onto the brackets, one per each side of the driving ram. Weights for treatments three through five will be added in increments of 220 additional N, or 110 N per bracket. This will be done until a total of 890 N has been added to the driving ram for treatment type five. From the five treatment types, the average value of depth per blow for each treatment type can be calculated. The averaged values can be used toward the regression analysis. The primary regression model of interest will be the weight added (X) versus penetration depth (Y). This model will allow for the determination of the impact of weight added to the driving ram on the driving depth of the vehicle. Additional supporting statistical analysis can be expounded upon as appropriate.

6. CONCLUSION

There are a vast number of locations in sub-Saharan Africa where increased population is expected over the coming years and where the water table is within the anticipated depth capabilities of the well-driver PUP. In locations where the number of drilling rigs are reduced and where labor is scarce, the well-driver PUP could be used to install shallow tube wells. This has the potential to improve the availability, quality, and accessibility of water in locations lacking sufficient water supplies. In order to increase the driving depth of the vehicle, increased weight was added to the driving ram channel. This was done by designing a ram weight bracket that would allow for incremental weight to be added to the driving ram on each side. The design of the weight bracket was conducted in Fusion 360. The FEA results yielded little to no deformation and minimal stresses. The bracket was designed to add to the ram an additional 890 N, or 445 N per side of the driving ram. Each bracket was designed with a safety factor of 15 in the event that additional weight beyond the intended 890 N might be desired in the future. The weight brackets and plates were fabricated following the total FEA. Moving forward, future efforts will be to conduct field testing, utilizing wooden fence posts. A regression model of the weight added (X) versus penetration depth (Y) will be developed from this testing. This model will allow for the determination of the impact of weight added to the driving ram on the driving depth of the vehicle. Additional supporting statistical analysis can be expounded upon as appropriate. Increasing the depth capabilities of the well-driver PUP has the potential to benefit the global community by moving an inexpensive option of sanitary water well installation closer to release in the developing world.

7. ACKNOWLEDGMENTS

The authors would like to thank Scott Brand for his assistance in the fabrication of the final ram weight bracket design and Mr. Elvis Kan-uge for providing cultural insight and revisions regarding the initial work. This research was a recipient of the Community Service/Service-Learning Student Grant Program of Purdue University. The assistance of the Purdue University Agricultural & Biological Engineering Department is gratefully acknowledged for its support over the years with graduate teaching assistantships and faculty salaries.

8. ENHANCEMENT OF WELL-DRIVER PUP QUESTIONS

1. Explain the original limitations to the well-driver component being added to the PUP vehicle and how it could affect the overall handling performance of the unit while traveling.
2. Describe the trade-offs between simply adding weight to the driving ram versus installing a larger overall post driver unit on the PUP vehicle.
3. Why were the additional weight brackets added to the bottom of the well driver ram outside the driving channel and not elsewhere?
4. What was the limiting factor to the addition of further ram weight on the current device, and was that limit reached with the current modification?
5. Describe the ideal locations for well-driver PUP activity that are worthy of being candidates for tube well installation.

9. REFERENCES

Baldwin, G. L. (2022). *Improvements to the driving capabilities of a well-driver PUP (Purdue Utility Project) to install low-cost driven water wells* [Doctoral dissertation]. Purdue University.

Baldwin, G. L., & Stwalley, R. M., III. (2018). *An agricultural extension demonstration farm template & community development project* [Paper presentation]. 2018 ASABE Annual International Meeting, Detroit. https://doi:10.13031/aim.201800693.

Baldwin, G. L., & Stwalley, R. M., III. (2019). *Analysis of market assessment survey to help promote lake restoration of Lake Bosomtwe and increased livelihoods for small-holder farmers* [Paper presentation]. 2019 ASABE Annual International Meeting, Boston. https://doi:10.13031/aim.201901379.

de Graff, I. E., Sutanudjada, E. H., van Beek, L. P., & Biekens, M. F. (2015). A high-resolution global-scale groundwater model. *Hydrology and Earth System Sciences, 19*(2), 823–837. https://doi:10.5194/hess-19-823-2015.

Driscoll, F. G. (1986). *Groundwater and wells.* Johnson.

Fidler, M. D., McPheron, T. J., & Soloman, R. W. (2022). *Improvements to the well-driver PUP: A final report.* Purdue University [an unpublished capstone project].

Held, A. (2017). *U.N. says world population will reach 9.8 billion by 2050.* In the Two Way [Amy Held Blog]. National Public Radio.

Horn, Z. J. (2019). *Prototyping a well-driver PUP (Purdue Utility Project) to install low-cost driven water wells* [Master's thesis]. Purdue University. https://doi:10.25394/PGS.8038991.v1.

Horn, Z. J., & Stwalley, R. M., III. (2018). *Well-driver PUP* [Paper presentation]. ASABE Annual International Meeting, Detroit. https://doi:10.13031/aim.201801196.

Horn, Z. J., & Stwalley, R. M., III. (2019). *A low-cost mechanized tube well installer* [Paper presentation]. 2019 ASABE Annual International Meeting, Boston. https://doi:10.13031/aim.201901319.

Horn, Z. J., & Stwalley, R. M., III. (2020). Design and testing of a mechanized tube well installer. *Groundwater for Sustainable Development, 11*, 100442. https://doi:10.1016/j.gsd.2020.100442.

International Groundwater Resources Assessment Centre. (2017). *African groundwater portal.* https://www.un-igrac.org/special-project/africa-groundwater-portal.

Purdue Utility Project. (2021). *Where.* engineering.purdue.edu: https://engineering.purdue.edu/pup/where/.

Shaver Manufacturing. (2009). *Operator's manual for hydraulic post driver model HD-8 & HD-8-H.* http://www.shavermfg.com/media/uploads/HD8-Operator-Manual.pdf.

United Nations. (2019). *World population prospects 2019: Highlights.* Department of Economic and Social Affairs, Population Division, United Nations.

US Geological Service. (2016). *How much water is there on, in, and above the Earth?* https://water.usgs.gov/edu/earthhowmuch.html.

Wada, Y., Wisser, D., & Bierkens, M. F. (2014). Global modeling of withdrawal, allocation, and consumptive use of surface water and groundwater resources. *Earth Systems Dynamics, 5*, 15–40. https://doi:10.5194/esd-5-15-2014.

Slide 1

"Improving the Driving Capabilities of a Well-Driver PUP (Purdue Utility Project) to Install Low-Cost Driven Water Wells

2022 ASABE Annual International Meeting
Houston, Texas
July 17–20, 2022

Grace L. Baldwin – Ph.D. Candidate
Tyler J. McPheron – M.S. Student
Dr. Robert M. Stwalley III
Agricultural & Biological Engineering
Purdue University

1

Slide 2

Background

- *"A PUP is a three wheeled low-cost vehicle designed for use in developing countries (PUP, 2021)*

- *"A hydraulic post driver mated to a PUP has been designed to mechanize the process of installing driven water wells (Horn, 2019)."*

2

Slide 3

Background
PUP

- Predominately used in sub-Saharan African Countries (PUP, 2021).

- Typically build in-country with locally source materials and minimal tooling

Figure 1. Well-Driver PUP (Baldwin, 2022)

3

Slide 4

Background
Application

- Locations where the number of drilling rigs are reduced and labor is scarce

- Improve the availability, quality, and accessibility of water in locations lacking sufficient water supplies.

- Well-Driver should be:
  - Easily to maintain
  - Not require any formal well drilling and or geology experience
  - Used as a tool for economic development

4

Slide 5

Literature Review
Groundwater

- Global estimates indicate 30% of the Earth's entire freshwater resources are found within the ground (USGS, 2016)
- 68% are locked up in ice and glaciers (USGS, 2016)
- World wide, groundwater is the primary source for
  - Drinking
  - Agricultural activities
  - Industrial activities (Wada et al., 2014)

5

Slide 6

Literature Review
Well-Driver PUP Depth

- Demonstrated depths of up to 7.0 m

- Analytically possible to reach 15 m

- Demonstrating and increasing the current depth can increase the number of applicable locations worldwide. (Horn, 2019).

6

Slide 7

Literature Review
Global Depth to Groundwater

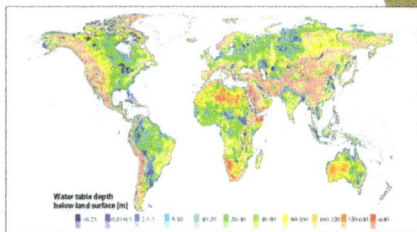

Figure 2. Water Table Depth Below Land Surface (de Graaf, et al., 2015).

7

Slide 8 (9)

Literature Review
Gaps within Prior Research

- Five test well installations
- Deepest well was 23 feet or 7 m
  - Utilized a submersible pump at 2.95 gpm
  - Continuous duty cycle pump, ran 8 hours a day
  - Supply 5,360 L of water to a community of 268 people (Horn, 2019).
- Didn't test any additional weight to driving ram

Objective:
- Increase driving ram weight
  - Based-on current ram weight, achievable depth of 50 ft is possible in favorable conditions (Driscoll, 1986; Horn, 2019).
- Increase depth of drive
- Increase the number of locations where the PUP could be a benefit.

9

Slide 9

Literature Review
Gaps within Prior Research

- Five test well installations
- Deepest well was 23 feet or 7 m
  - Utilized a submersible pump at 2.95 gpm
  - Continuous duty cycle pump, ran 8 hours a day
  - Supply 5,360 L of water to a community of 268 people (Horn, 2019).
- Didn't test any additional weight to driving ram

Objective:
- Increase driving ram weight
  - Based-on current ram weight, achievable depth of 50 ft is possible in favorable conditions (Driscoll, 1986; Horn, 2019).
- Increase depth of drive
- Increase the number of locations where the PUP could be a benefit.

9

Slide 10

Methodology

Increase the ram weight of the driving ram to achieve greater depths

- Design:
  - Brackets for the lower portion of the driving ram
  - Individual weight plates that can be added to brackets

- Model Bracket Design in Fusion 360
  - 3D modeling
  - Part drawings
  - Finite element analysis (FEA) in Fusion 360
- Fabricate Design

10

Methodology

- Effective weight of spring powered driving ram – 1600 N (Shaver, 2009)
- Static Weight of Ram – 1020 N
- Spring of the driving ram – 580 N

A 1.15 factor allows for testing the addition of up to 880 N
(Fidler, McPheron, Solomon, 2022)

Figure 4. Fusion 360 Unloaded Ram Weight Bracket Design (Fidler et al., 2022)
Ram Channel – (Grey), Plate Weld – (Yellow), Weight Plate Weld – (Blue), Weight Support Plate – (Red)

11

Methodology

- Each suitcase weight – 110 N, A36 Steel
- Possible addition of 440 N per side
- Total cumulative weight of 880 N
- Load case modeled:
  - Uniformly distributed
  - Dynamic load of 1110 N
  - Vertically applied to top weight plate
- Analysis included
  - Displacement, Safety factor, and Von Misses Stress

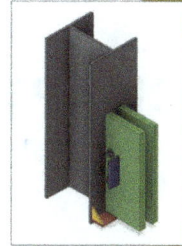

Figure 5. Fusion 360 Ram Weight Bracket Loaded with Weight Plates *(Green)* (Fidler et al., 2022)

12

Results
Safety Factor

- Intentionally over-engineered to provide substantial support
- Ultimate load of 16,690 N & a Dynamic load of 1110 N
- F.S. of 15
- High F.S. allows for increased weights beyond 880 N, if required in the future.
- Minimal displacement ($<1.2 \times 10^{-4}$ cm)

Figure 6. Safety Factor Fusion 360 Analysis (Fidler et al., 2022)

13

Results
Fabrication

Figure 8. Fabricated Ram Weight Bracket Design Loaded with Weight Plate (Fidler et al., 2022)

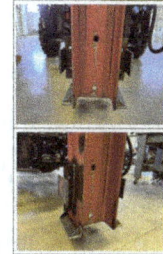

Figure 7. Fabricated Unloaded Ram Weight Bracket Design (Fidler et al., 2022)

14

Future Testing

- Install wooden fence posts with beveled ends
  - Minimal cost compared to real well points
  - Five treatments, 4 posts per driven weight used
    - Added in increments of 220 N, 110 N per bracket
- Develop regression model from experiment
- Additional supporting statistical analysis conducted as appropriate

15

Conclusion

- Vast number of locations where the Well-Driver PUP could be a benefit
- Best suited in locations where drilling rigs are reduced, and labor is scare
- Potential to improve water availability, quality, and accessibility in locations lacking adequate supplies
- Additional weight is needed to reach deeper depths
- The additions of ram weight brackets, and weight plates, allow for weight to be added
- Design, modeling, and analysis were completed in Fusion 360

16

Conclusion

- Each bracket allows for the addition of 440 N per side.
- F.S. of 15 to allow for weight beyond the 880 N if needed in the future
- Little to no deflection from the analysis
- Weight brackets and weight plates were fabricated
- Weight brackets were welded to the drive ram channel
- Future in field testing is planned
- Development of a regression model from field testing is planned
  - Weight Added (X) vs Penetration Depth (Y)

17

Contact Us

Dr. Robert M. Stwalley
rms3@purdue.edu

Grace L. Baldwin
baldwing@purdue.edu

Agricultural & Biological Engineering
Purdue University, West Lafayette IN USA

CHAPTER 4. REGRESSION OF DRIVING CAPABILITY MODEL

Theoretical depth limitations to the well-driver PUP suggested that enhancement of this constraint might broaden the useful range of the machine in providing tube well installations. Brackets were designed to add additional suitcase-type blocks of steel to the bottom of the ram. An experimental study was designed and undertaken to determine the effect of additional weight increments on the potential driving depth of the unit. Four treatment combinations of additional ram weight ranging from 0 to 720 N were tested using sharpened wooden fence posts. Additional weight, drop height, and soil moisture content were used as independent variables. The regression model was developed using 48 unique post drives of 10 blows each. Measurements were taken after each machine stroke. All data were normal, and there were no outliers requiring removal via Student's residual test. The resulting model was statistically significant and could be used in designing a new prototype system for tube well installation. This paper was originally presented at the 2023 annual international meeting of the American Society of Agricultural and Biological Engineers. The poster presented at the meeting is also included.

DESIGN AND DEVELOPMENT OF A PRELIMINARY REGRESSION MODEL TO DETERMINE THE DRIVING CAPABILITIES OF THE WELL-DRIVER PUP

GRACE L. BALDWIN KAN-UGE, TYLER J. MCPHERON (PURDUE UNIVERSITY AGRICULTURAL & BIOLOGICAL ENGINEERING), AND ROBERT M. STWALLEY III

ABSTRACT

There are a vast number of locations in sub-Saharan Africa where increased population is expected over the coming years and the water table is within the anticipated depth capabilities of the Purdue well-driver PUP. The well-driver PUP could be used to install shallow tube wells, especially in locations where the number of drilling rigs is reduced and labor is scarce. This apparatus has the potential to improve the availability, quality, and accessibility of water in locations lacking sufficient water supplies. In order to increase the driving depth of the vehicle, increased weight was added to the driving ram channel. This was done by designing a ram weight bracket that would allow for incremental weight to be added to each side of the driving ram. An experiment was conducted to determine the driving capabilities of the vehicle through a series of wooden fence post installations using these new weight additions. Four driving ram weight treatments and three different ram weight drop heights were tested. This experiment allowed for a regression model to be developed predicting the impact of weight added to the driving ram on the driving depth of the vehicle. This paper will go into further detail of the experimental results and the developed regression model and their implications to increase the driving depth capabilities of the well-driver PUP.

KEYWORDS: Africa, groundwater, purdue utility platform, tube well, international development, sub-saharan, water, water resources, well driver, wells

1. INTRODUCTION

Globally, 1.8 billion people use an unimproved source of drinking water with no protection against contamination from feces. Combined with safe water and good hygiene, improved sanitation could potentially prevent around 842,000 deaths each year. This initiative seeks to provide progress toward addressing the sixth United

Nations sustainable development goal, "to ensure availability and sustainable management of water and sanitation for all," which will help drive progress across many other sustainable development goals (United Nations, 2019). This research seeks to increase access to subsurface water by improving the driving capabilities of the well-driver Purdue Utility Platform (PUP) vehicle. Increasing the vehicle's achievable depth capability will increase the number of potential locations throughout the world where the vehicle could be used to access groundwater for those with limited to no current water access. Freshwater access has been shown to be critical in community health through the improvement of water, sanitation, and hygiene (WASH) elements (Baldwin & Stwalley, 2022c; Baldwin & Stwalley, 2020; Baldwin & Stwalley, 2019; Baldwin, 2019; Baldwin & Stwalley, 2018; World Health Organization, 2016) with the potential for local economic development through irrigated agriculture (Baldwin & Stwalley, 2022b; Baldwin & Stwalley, 2021). Implementation and dissemination of this vehicle technology can improve access to safe water in developing countries for drinking and domestic purposes.

A hydraulic post driver mated to a PUP has been designed to mechanize the process of installing driven water wells, also called the Well-Driver PUP (Baldwin Kan-uge et al., 2023; Horn & Stwalley, 2020; Horn, 2019). The well-driver PUP could potentially reduce dependency on manual labor for driven well installation in developing countries and keep laborers safer (Horn, 2019). A standard PUP is a three-wheeled low-cost utility vehicle designed for use in developing countries. These vehicles are typically built in-country with only locally sourced available materials and minimal tooling, predominantly for sub-Saharan African countries (Purdue Utility Project, 2021). The well-driver PUP is pictured in Figure 4.1. Standard PUPS are primarily used for transportation purposes in areas where normal vehicles are not appropriate due to the severe terrain and lack of road accessibility. Multiple capstone teams from within the Purdue Agricultural & Biological Engineering program have contributed to the success of this project (Fidler et al., 2022), and the details of the well-driver design can be found in Baldwin & Stwalley (2022a), Horn & Stwalley (2020), Horn & Stwalley (2019), and Horn & Stwalley (2018).

In locations where the number of drilling rigs is reduced and where labor is scarce, the well-driver PUP vehicle could be used as an instrument of economic development, providing microbusiness development opportunities focused on well installation. Installing low-cost driven wells would improve the availability,

FIGURE 4.1. Well-driver PUP (Baldwin Kan-uge et al. 2023)

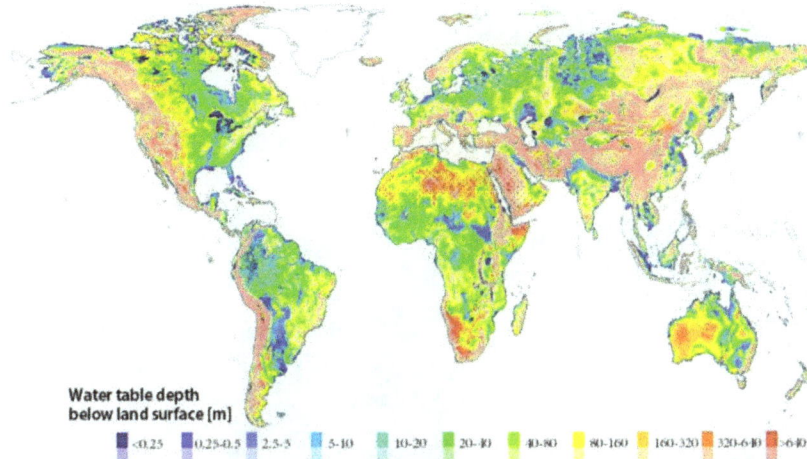

Water table depth below land surface [m]

| <0.25 | 0.25-0.5 | 2.5-5 | 5-10 | 10-20 | 20-40 | 40-80 | 80-160 | 160-320 | 320-640 | >640 |

FIGURE 4.2. Water table depth below land surface (de Graff et al. 2015). The water table is most shallow in eastern North and South America; central Africa; and central Europe. It is deepest in western North and South America; Saharan and south Africa; southern Europe; the Arabian Peninsula; eastern Asia; and western Australia.

quality, and accessibility of water in locations lacking sufficient water supplies. In order to be successful, the well-driver PUP operation and components must remain low-cost, easy to maintain, and based on locally accessible materials as much as possible. It is intended that operation of the vehicle should require no formal training in well drilling and/or geology. In locations with an appropriate water table depth, this vehicle could decrease both the wait time to install wells and the final cost of installation (Horn, 2019).

2. LITERATURE REVIEW

2.1. POTENTIAL IMPACT OF THE WELL-DRIVER PUP IN THE DEVELOPING WORLD

Global estimates indicate that 30% of Earth's entire freshwater resources are found in the ground, while 68% is locked up in ice and glaciers (United States Geological Survey, 2016). Groundwater serves as the world's largest source of freshwater and plays a critical role in meeting the needs of people around the world (de Graaf et al., 2015). Groundwater is the primary source of drinking water and supplies water for agricultural and industrial activities worldwide (Wada et al., 2014). The depth to the water table varies throughout the world. Demonstrating the well-driver PUP's capability of achieving depths of up to 15 m would increase the number of locations throughout the world where the well-driver PUP could be utilized to access water for individuals who currently have little to no access. A high-resolution global-scale groundwater model was developed by de Graaf et al. (2015). They model they created, highlighted from their model simulations the current water table depth on a global scale. Figure 4.2 provides insight as to the water table depth below the land surface throughout the world.

Worldwide, there are many locations where the water table depth is within 15 m, specifically in the 10–20 m range. Many of these locations are in sub-Saharan Africa, South America, northern India, Asia, and parts of the Asia Pacific Islands. These locations demonstrate potential places where the well-driver PUP could be utilized if sufficient driving depth can be demonstrated on a repeatable basis. Through testing, Horn (2019) demonstrated that the well-driver PUP could achieve depths of up to 7.0 m. Horn (2019) analytically determined that the vehicle should be capable of achieving depths of up to 15 m without significant changes (Driscoll, 1986; Horn, 2019).

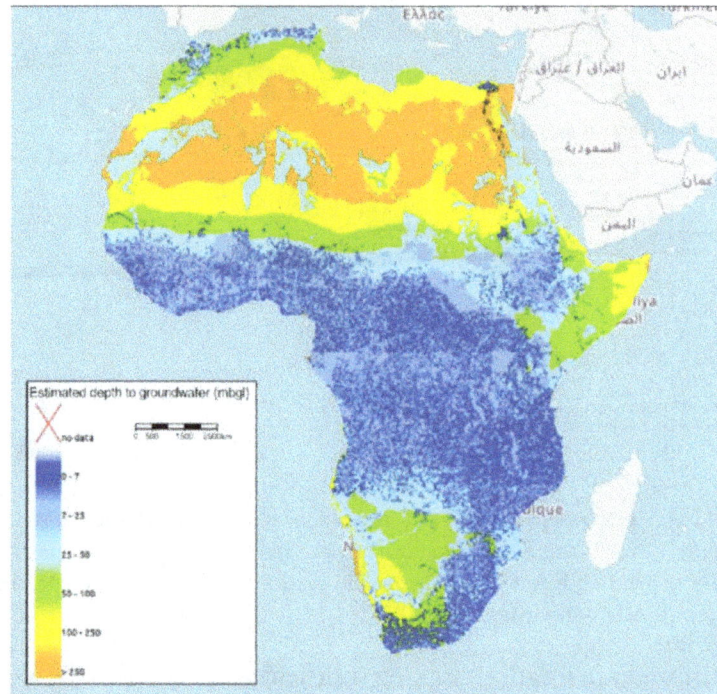

FIGURE 4.3. Estimated depth to groundwater (*mbgl*) and transboundary aquifer of Africa (IGRAC, 2017). The groundwater is deepest in the Sahara and southwestern portions of the continent. It is shallower in sub-Saharan Africa.

According to the United Nations, approximately 14.5% of the world's population is located in sub-Saharan Africa (United Nations, 2019). The population of sub-Saharan Africa alone is projected to double by 2050 (99%) (United Nations, 2019). Based on data where the well-driver PUP could impact the largest number of people, specific consideration should be given to locations in sub-Saharan Africa. The International Groundwater Resources Assessment Centre (IGRAC) and the UNESCO International Hydrological Programme have mapped the transboundary aquifers in Africa (IGRAC, 2017). IGRAC, in collaboration with the British Geological Survey and the University College London, has developed maps to quantify the groundwater resources in Africa. Their results are highlighted in Figure 4.3 and provide a high-resolution look at the depth to the water table in Africa. When specifically looking at sub-Saharan Africa, the approximate depth to the groundwater is predominantly within 25 *m* or less, followed by those in the 25–50 *m* range. This demonstrates the vast number of locations in sub-Saharan Africa where the well-driver PUP could access groundwater and has great potential to make a substantial positive impact for those with little to no water access.

3. METHODOLOGY

In order to increase the driving depth of the vehicle's well driver, increased weight was added to the driving ram channel (Fidler et al. 2022). This was done by designing a ram weight bracket that would allow for incremental weight to be added to each side of the driving ram (Baldwin Kan-uge, 2023). An experiment was conducted to determine the driving capabilities of the well driver through a series of wooden fence post installations using these new weight additions. This experiment allowed for a regression model to be developed of the impact of weight added, ram drop height, soil moisture content, and the driving depth of the vehicle. Horn (2019) first demonstrated the driving capabilities of the well-driver through the installation of

TABLE 4.1. *Treatment type by weight added for post driving experiment*

TREATMENT NUMBER	DRIVING RAM TOTAL WEIGHT (N)	WEIGHT ADDED (N)	NUMBER OF WEIGHT PLATES USED	NUMBER OF POSTS DRIVEN PER TREATMENT
1	1,076	0	0	12
2	1,317	240	6	12
3	1,557	480	12	12
4	1,797	720	18	12

wooden fence posts. During the well driving process, the well point remains in the ground once driven. The wooden fence posts can be pulled up using a tractor upon completion. Utilizing wooden fence posts allowed for the experimental costs to be significantly less expensive than if a formal well point, a coupler, a piping section, and a drive cap were used for each sample in the experiment. The wooden fence posts used were 4 *in* × 4 *in* × 8 *ft* (10 *cm* × 10 *cm* × 244 *cm*). The weight of each fence post was approximately 30 *lb*$_f$ (133 *N*). Each wooden post was beveled at the end to mimic a well point. Wooden fence posts were premarked at 1 and 0.5 *ft* increments for easy measurement during driving. This practice was also followed by Horn (2019). The experiment was carried out in rural Montgomery County, Indiana. Four driving ram weight treatments were used: 0, 240, 480, and 720 *N*. Three ram drop heights were used: 0.254, 0.508, and 0.762 *m*. Twelve different blocking locations were selected in the farm site. For each combination of weight added and drop height, four wooden posts were driven. For each post, a total of 10 blows was applied. Table 4.1 displays the weight amount added and the number of weight plates per treatment.

For each set of 12 posts, the block location, weight added, and ram drop height were varied. The first round of 12 posts driven consisted of only the ram raw weight and the welded brackets (1076 *N*). The next set of 12 posts included the weight addition of 240 *N*. This was followed by another 12 posts driven using 480 *N* of additional weight and a final 12 posts driven with weight added of 720 *N*. The driving data collected for each treatment included the driving ram stroke count and the aboveground height of the wooden post at each stroke. The vehicle ram cylinder pressure gage readings remained predominantly constant throughout the experiment. After obtaining these data, the depth belowground of each wooden post was calculated. Soil moisture samples were obtained for each post driven so that a corresponding average moisture content could be calculated. ASTM D2216–98 standard was used to determine the moisture content of the soil samples (American Society for Testing and Materials, 1998). Statistical analysis and regression modeling was conducted in Excel and R.

4. RESULTS

The penetration depth for each post was calculated from the experimental field data collected. The data used to develop a preliminary multiple linear regression (MLR) model included the penetration depth (Y), weight added (X_1), drop height (X_2), and soil moisture content (X_3).

$$Y_i = \beta_0 + \beta_1 X_1 + \beta_2 X_2 + \beta_3 X_3 + \varepsilon_i \tag{1}$$

The MLR model could be used to determine if the independent variables, weight added, drop height, and soil moisture content significantly predicted the penetration depth of the well-driver PUP. Diagnostic analysis was conducted to verify that the model data followed a normal distribution; the Shapiro-Wilk test for

TABLE 4.2. *Evaluation of potential outliers in the post driving experiment*

	STUDENT TEST RESIDUALS	HAT VALUES
5	−0.3578902	0.13419272
17	−0.8518417	0.13822383
33	−2.5131137	0.09748767
48	3.0793826	0.09136578

TABLE 4.3. *Regression statistics from the post driving experiment*

Multiple R	0.88
R^2	0.77
Adjusted R^2	0.76
Standard Error	1.36
Observations	48

TABLE 4.4. *ANOVA results from the post driving experiment*

	Df	SS	MS	F	Significance F
Regression	3	272.33	90.78	49.41	3.88E-14
Residual	44	80.84	1.84		
Total	47	353.17			

normality was conducted in R. The data was found to follow a normal distribution (W= 0.985, p-value = 0.777). The influence function was used in R to determine if any outliers existed in the data due to their X or Y values. These values are provided in Table 4.2.

No outliers were found due to their Y or X values. The student test residuals were all less than the critical value (t(0.999, 43) = 3.29), indicating no outliers with respect to Y. The Hat values were all found to be less than $2\hbar$ = 0.167, indicating no outliers with respect to X. The variance inflation factor for each variable was calculated: V_1=1.05, V_2=1.01, and V_3=1.06. Since the variance inflation factor values are approximately 1 each, there was no concern of multicollinearity between the variables.

From conducting regression analysis, the beta values were estimated, and corresponding p-values were obtained. Tables 4.3–4.5 provide sample results from the analyses. From the regression statistics Table 4.3, the adjusted R^2 = 0.76. This value is indicative that the set of predictor variables did a good job of explaining the response variable penetration depth. The standard error for the model was also small at only 1.36. Tables 4.4 and 4.5 demonstrate that from the ANOVA analysis the regression model containing all variables is significant ($F_{(3, 44)}$ = [49.41], p < 0.01).

TABLE 4.5. *ANOVA results from the post driving experiment continued*

	COEFFICIENTS	STANDARD ERROR	T-STATISTIC	P-VALUE	LOWER 95.0%	UPPER 95.0%
Intercept	1.448	0.97	1.50	1.41E-01	−0.50	3.39
X1, Weight added	0.005	0	6.13	2.16E-07	0.00	0.01
X2, Drop height	0.241	0.02	9.98	7.12E-13	0.19	0.29
X3, Soil moisture content	0.106	0.04	2.97	4.82E-03	0.03	0.18

Table 4.5 shows the model coefficient estimates, where the predictor variables were found to all be significant at $p < 0.01$. The overall regression model was statistically significant (Adjusted R^2 = [0.76], $F_{(3, 44)}$ = [49.41], p = [3.88E-14]). The fitted regression model was determined to be

$$Y = 1.448 + 0.005X1 + 0.241X2 + 0.106X3 \qquad (2)$$

5. CONCLUSION

There are many locations in the world, particularly in sub-Saharan Africa, where water access is a concern and the use of the well-driver PUP could be a benefit to local communities. Increasing the weight of the driving ram has the potential to improve the driving depth capabilities of the vehicle. An experiment consisting of a series of wooden fence post installations using these new weight additions was conducted. This experiment allowed for a regression model to be developed predicting the impact of weight added to the driving ram, the drop height of the ram, and the soil moisture content on the driving depth of the vehicle. The preliminary MLR model included the penetration depth (Y), weight added (X_1), drop height (X_2), and soil moisture content (X_3). No outliers within the data were found. Multicollinearity between variables was determined to not be a concern. Regression analysis was conducted in Excel and R. The model coefficient estimates were determined, and the predictor variables were all found to be significant at $p < 0.01$. The preliminary regression model was statistically significant (Adjusted R^2 = [0.76], $F_{(3, 44)}$ = [49.41], p = [3.88E-14]). The final regression model was shown as Equation 2. This experiment allowed for a preliminary MLR model to be developed for predicting the impact of the predictor variables on the penetration depth of the well-driver PUP. This work will have impact on the overall effects of increasing the driving depth of the well-driver PUP.

6. ACKNOWLEDGMENTS

The authors would like to thank Mr. Robert M. Stwalley Jr. for generously allowing the experimental testing to take place on his farm. This research was a recipient of the Community Service/Service-Learning Student Grant Program of Purdue University. The assistance of the Purdue University Agricultural & Biological Engineering Department is gratefully acknowledged for its support over the years with graduate teaching assistantships and faculty salaries.

7. REGRESSION MODEL OF WELL-DRIVER PUP QUESTIONS

1. If the depth per driver blow is the dependent variable in this study, how many total individual data points were collected for this study?

2. If a fully blocked study was conducted using three different drop heights, how many repetitions were performed for each drop/weight combination?

3. Why were the additional weight brackets added to the bottom of the well-driver ram outside the driving channel and not elsewhere?

4. Explain why the normality of the data and the lack of outliers are important to the overall statistical analysis?

5. Describe how the measurement units affect the coefficients in the final regression equations (i.e., the coefficient for weight added is small, and the coefficient for drop height is large).

8. REFERENCES

American Society for Testing and Materials. (2019). *Standard test materials for laboratory determination of water (moisture) content of soil and rock by mass (Standard No. D2216–98).* ASTM. https://doi:10.1520/D2216–98.

Baldwin, G. L. (2019). *Development of design criteria and options for promoting lake restoration of Lake Bosomtwe and improved livelihoods for small-holder farmers near Lake Bosomtwe-Ghana, West Africa* [Master's thesis]. Purdue University.

Baldwin, G. L., & Stwalley, R. M., III. (2018). *An agricultural extension demonstration farm template & community development project* [Paper presentation]. ASABE 2018 AIM, Detroit. St. Joseph. https://doi:10.13031/aim.201800693.

Baldwin, G. L., & Stwalley, R. M., III. (2019). *Analysis of market assessment survey to help promote lake restoration of Lake Bosomtwe and increased livelihoods for small-holder farmers* [Paper presentation]. ASABE 2019 AIM, Boston. https://doi:10.13031/aim.201901379.

Baldwin, G. L., & Stwalley, R. M., III. (2020). *Promoting restoration of Lake Bosomtwe through spatial analysis of existing water, sanitation, and hygiene (WASH) sources in Ghana, West Africa* [Paper presentation]. ASABE 2020 AIM, Pasadena. https://doi:10.13031/aim.202000589.

Baldwin, G. L., & Stwalley, R. M., III. (2021). *An economic analysis: The scale-up of irrigation systems in Ghana, West Africa* [Paper presentation]. ASABE Annual International Meeting, Pasadena. https://doi:10.13031/aim.212100009.

Baldwin, G. L., & Stwalley, R. M., III. (2022a). *Improving the driving capabilities of a well-driver PUP (Purdue Utility Project) to install low-cost driven water wells* [Paper presentation]. ASABE Annual International Meeting, Houston. https://doi:10.13031/aim.202200203.

Baldwin, G. L., & Stwalley, R. M., III. (2022b). Opportunities for the scale-up of irrigation systems in Ghana, West Africa. *Sustainability, 14,* 8716. https://doi:10.3390/su14148716.

Baldwin, G. L., & Stwalley, R. M., III. (2022c). A review of freshwater programming and access options in Ghana, West Africa. *African Journal of Water Conservation and Sustainability, 10*(4), 1–19.

Baldwin Kan-uge, G. L. (2023). *Improvements to the driving capabilities of a well-driver PUP (Purdue Utility Project) to install low-cost driven water wells* [Doctoral dissertation]. Purdue University.

Baldwin Kan-uge, G. L., McPheron, T. J., Horn, Z. J., & Stwalley, R. M., III. (2023). Cost-effective, sanitary shallow water wells for agriculture and small communities using mechanized tube well installation. In J. Tarhouni, *Groundwater: New advances and challenges.* Intech Open. https://doi:10.5772/intechopen.109576.

de Graff, I. E., Sutanudjada, E. H., van Beek, L. P., & Biekens, M. F. (2015). A high-resolution global-scale groundwater model. *Hydrology and Earth System Sciences, 19*(2), 823–837. https://doi:10.5194/hess-19-823-2015.

Driscoll, F. G. (1986). *Groundwater and Wells.* Johnson.

Fidler, M. D., McPheron, T. J., & Soloman, R. W. (2022). *Improvements to the well-driver PUP: A final report.* Purdue University [an unpublished capstone project].

Horn, Z. J. (2019). *Prototyping a well-driver PUP (Purdue Utility Project) to install low-cost driven water wells* [Master's thesis]. Purdue University. https://doi:10.25394/PGS.8038991.v1.

Horn, Z. J., & Stwalley, R. M., III. (2018). *Well-driver PUP* [Paper presentation]. ASABE Annual International Meeting, Detroit. https://doi:10.13031/aim.201801196.

Horn, Z. J., & Stwalley, R. M., III. (2019). *A low-cost mechanized tube well installer* [Paper presentation]. ASABE 2019 AIM, Boston. https://doi:10.13031/aim.201901319.

Horn, Z. J., & Stwalley, R. M., III. (2020). Design and testing of a mechanized tube well installer. *Groundwater for Sustainable Development*, *11*, 100442. https://doi:10.1016/j.gsd.2020.100442.

IGRAC. (2017). *Africa groundwater portal*. International Groundwater Resources Assessment Centre of the United Nations. https://www.un-igrac.org/special-project/africa-groundwater-portal.

Purdue Utility Project. (2021). *Where*. https://engineering.purdue.edu/pup/where/.

United Nations. (2019). *World population prospects 2019: Highlights*. Department of Economic and Social Affairs, Population Division. https://population.un.org/wpp/Publications/Files/wpp2019_10KeyFindings.pdf.

United States Geological Survey. (2016). *How much water is there on, in, and above the Earth?* https://water.usgs.gov/edu/earthhowmuch.html.

Wada, Y., Wisser, D., & Bierkens, M. F. (2014). Global modeling of withdrawal, allocation, and consumptive use of surface water and groundwater resources. *Earth Systems Dynamics*, *5*, 15–40. https://doi:10.5194/esd-5-15-2014.

World Health Organization. (2016). *What is the minimum quantity of water needed?* https://www.who.int/water_sanitation_health/emergencies/qa/emergencies_qa5/en/.

PURDUE UNIVERSITY®

Agricultural and Biological Engineering

Design and development of a preliminary regression model to determine the driving capabilities of the Well-Driver PUP

Grace L. Baldwin Kan-uge, Tyler J. McPheron, & Robert M. Stwalley III

Introduction

- A PUP is a three wheeled low-cost utility vehicle designed for use in developing countries. [1]
- The Well-Driver PUP is a traditional PUP with a hydraulic post driver mated to it (Figure 1). [2]
- This vehicle could be used to install shallow tube wells, particularly in locations where labor and traditional drilling rigs are scarce. [2]
- An experiment was conducted to determine the driving capabilities of the vehicle through a series of wooden fence post installations and varied weight additions. [3]

Potential Impact

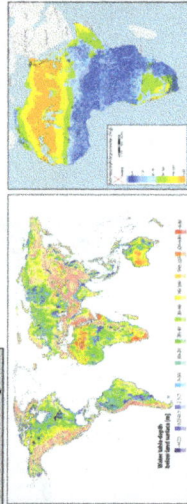

Water Table Depth Below Land Surface [4].

Estimated depth to groundwater (mbgl) of Africa [5].

- Groundwater is the primary source of drinking water and supplies water for agricultural and industrial activities worldwide [6].
- Worldwide, there are many locations where the water table depth is within 15 meters, specifically in the 10-20 meter range [3].
- The world population is expected to increase over the coming years, particularly in Sub-Saharan Africa [7].
- There are many locations in Sub-Saharan Africa where the water table is within the anticipated depth capabilities of the Purdue Well-Driver PUP [3].
- The Well-Driver PUP has the potential to improve the availability, quality, and accessibility of water in locations lacking sufficient water supplies [2,3].

Methods

- Increased vehicle driving depth through weight additions to ram channel [8].
- Determined the driving capabilities through a series of wooden fence post installations [3].
- Created a regression model from the experiment.
- Model components included the impact of weight added, ram drop height, soil moisture content, on the driving depth of the vehicle.
- Utilized wooden fence posts to minimize cost and experimental waste [3].

Well-Driver PUP [2].

Methods Continued

- Each wooden post was $4 \ in \ x \ 4 \ in \ x \ 8 \ ft$ and weighted $30 \ lb_f (133 \ N)$ [3].
- Twelve different blocking locations were selected within the farm site.
- Four driving ram weight treatments were used and three ram drop heights: 0.254, 0.508, and 0.762 m.
- Four wooden posts driven per combination.
- A total of 10 blows was applied per post.
- Block location, weight added, and ram drop height were varied for each post.

Treatment Type by Weight Added for Post Experiment [3].

Treatment Number	Driving Ram Total Weight (N)	Weight Added (N)	Number of Weight Plates Used	Number of Posts Driven Per Treatment
1	1076	0	0	12
2	1317	240	6	12
3	1557	480	12	12
4	1797	720	18	12

Note: Original static raw weight of the driving ram was 1023 N.

- Data collected: driving ram stroke count, above ground height of post, penetration depth, and soil samples per post location.
- Statistical analysis and regression modeling conducted in Excel and R.
- Created preliminary multiple linear regression (MLR) model of penetration depth (Y), weight added (X₁), drop height (X₂), and soil moisture content (X₃) [3].

Results

Regression Statistics [3].

Multiple R	0.88
R²	0.77
Adjusted R²	0.76
Standard Error	1.36
Observations	48

ANOVA Results [3].

	df	SS	MS	F	Significance F
Regression	3	272.33	90.78	49.41	3.88E-14
Residual	44	80.84	1.84		
Total	47	353.17			

- Diagnostic analysis to verify normality (W= 0.985, p-value =0.777).
- Utilized influence R function and found no outliers with respect to Y or X [3].
- Student Residuals all less than the critical value (t(0.999, 43) = 3.29).
- Hat values all less than 2h = 0.167.
- No multicollinearity concern between the variables.
- Each near approximately one (V1=1.05, V2=1.01, and V3=1.06).

ANOVA Results Continued [3].

	Coefficients	Standard Error	t-statistic	P-value	Lower 95.0%	Upper 95.0%
Intercept	1.448	0.97	1.50	1.41E-01	-0.50	3.39
X1 - Weight Added	0.005	0.00	6.13	2.16E-07	0.00	0.01
X2 - Drop Height	0.241	0.02	9.98	7.12E-13	0.19	0.29
X3 - Soil Moisture Content	0.106	0.04	2.97	4.82E-03	0.03	0.18

- Post-regression analysis, beta values estimated with corresponding p-values.
- Adjusted $R^2 = 0.76$. Indicative the set of predictor variables are doing a good job of explaining the response variable penetration depth [3].
- Minimal standard error for the model (1.36).
- ANOVA analysis results indicate the regression model containing all variables is significant ($F(3, 44) = [49.41], p < 0.01$).
- Model predictor variables all significant at ($p < 0.01$).
- Overall regression model statistically significant (Adjusted R2 = [0.76], F(3, 44) = [49.41], p = [3.88E-14]).
- Preliminary regression model:

$$Y = 1.448 + 0.005X_1 + 0.241X_2 + 0.106X_3$$

Conclusion

- A series of wooden fence post installations was conducted [3].
- Preliminary MLR model created of the penetration depth (Y), weight added (X1), drop height (X2), and soil moisture content (X3).
- The model coefficient estimates were determined, and the variables all found to be significant (p < 0.01).
- The model was statistically significant (Adjusted R2 = [0.76], F(3, 44) = [49.41], p = [3.88E-14]).
- The experiment led to a preliminary MLR model to be developed of the impact of the predictor variables on the penetration depth of the Well-Driver PUP [3].

References:
1. Purdue Utility Project (PUP). (2021). *iPhow.* Retrieved April 30, 2022, from engineering.purdue.edu: https://engineering.purdue.edu/pup/where/
2. Baldwin Kan-uge, G.L., McPheron, T. J., Horn, Z. J., & Stwalley III, R. M. (2023). Cost-effective, sanitary, shallow water wells for agriculture and small communities using mechanized tube well installation. in J. Tathoum, *Groundwater - New Advances and Challenges.* London, UK: Intech Open. doi:10.5772/intechopen.109576
3. Baldwin Kan-uge, G. L. (2023). *Improvements to the driving capabilities of a well-driver PUP (Purdue Utility Project) to install low-cost driven water wells.* Agricultural & Biological Engineering Doctoral Thesis (Unpublished). West Lafayette: Purdue University.
4. de Graff, I. E., Sutanudjaja, E. H., van Beek, L. P., & Bierkens, M. F. (2015). A high-resolution global-scale groundwater model. *Hydrology and Earth System Sciences*, 19(2), 823-837. doi:10.5194/hess-19-823-2015
5. BGRAC. (2017). *Africa groundwater portal.* Retrieved December 15, 2021, from International Groundwater Resources Assessment Centre of the United Nations (IGRAC): https://www.un-igrac.org/special-project/africa-groundwater-portal
6. Wada, Y., Wisser, D., & Bierkens, M. F. (2014). Global modeling of withdrawal, allocation, and consumptive use of surface water and groundwater resources. *Earth Systems Dynamics*, 5, 15-40. doi:10.5194/esd-5-15-2014
7. United Nations Department of Economic and Social Affairs, Population Division. (2019). *World Population Prospects 2019: Highlights.* New York, UN. Retrieved December 15, 2020, from https://population.un.org/wpp/Publications/Files/wpp2019_10KeyFindings.pdf
8. Feller, M. D., McPheron, T. J. & Solomon, R. W. (2022). *Improvements to the well-driver PUP: a final report.* West Lafayette: Purdue University (an unpublished capstone project)

CHAPTER 5. NOVEL DRIVEN WELL PIPE COUPLER

Tube well pipe stacks installed with the driver mechanism are subjected to significantly higher impact forces than stacks driven manually. Drive couplings used between pipe sections are different than standard pipe couplers due to their design to ensure that pipe sections butt end to end rather than simply being supported by the mating threads. This system works reasonably well for tube wells installed by either hammer blows or dead-drop weights, but it is insufficient for use with the hydraulic driving system. Driving pipe stacks to deeper depths tends to destroy the connections at the couplers even when using the recommended drive couplings. These failures ruin the entire installation. Purdue researchers have developed a new and unique press fit pipe coupling specifically designed to survive the impact blows from the driving apparatus. One side of the coupler is welded to a pipe section, and the other has side an interference fit and is driven into place. This novel approach succeeded in eliminating the pipe thread failures in the tube well stacks that could cause the overall failure of the installation. The lack of integrity in the well pipe stack can create a contaminate entry pathway into the well and destroy its hygienic nature. The novel couplers were approved for use by the Indiana Department of Natural Resources Water Division, were tested in the field, and performed without a problem. This paper was originally presented at the 2024 annual international meeting of the American Society of Agricultural and Biological Engineers. The poster presented at the meeting is also included.

PROOF-OF-CONCEPT TESTING FOR A NOVEL WELL PIPE COUPLER USED IN SHALLOW TUBE WELLS

TYLER J. MCPHERON (PURDUE UNIVERSITY AGRICULTURAL & BIOLOGICAL ENGINEERING), GRACE L. BALDWIN KAN-UGE, AND ROBERT M. STWALLEY III

ABSTRACT

A series of ultimately successful tube well installations were completed in 2023 by the well-driver Purdue Utility Platform (PUP) team using a modified post driver to drive 2 *in* nominal diameter well pipes into the soil, using a ram to provide an impact force. Alignment issues occurred with the assembly of the pipe stack throughout experimental well installation tests. It was discovered that due to the dimensional inconsistencies of the pipe thread, joining pipes did not always contact the mating surfaces on the interior of specially made threaded pipe couplers used for water well construction. Without the ability to transfer load to the mating surfaces inside the coupler, the driving forces moved through the national pipe threads of the junction. Damaged threads degraded the structural rigidity of the pipe stack and were the primary contributing factor to bent pipe installation failures. Weakened pipe stacks tended to bend. To correct this problem, a prototype interference-fit pipe coupler was designed and machined from carbon steel drawn-over mandrel tubing. This coupler was designed with a tight tolerance interference fit on one end and a slip fit on the other end. The interfering end was machined-undersized 0.076 *mm* with a ±0.051 *mm* tolerance, while the slip-fit end was machined-oversized up to 0.127 *mm*. Each coupler was manually machined using a lathe and a boring bar. The well drive point and each pipe section were inserted onto the slip-fit end of the coupler and welded solid before going to the field for a well installation. This left the interference fit end of the coupler exposed and available for coupling. At the well installation site, the joining pipe was inserted minimally into the undersized end of the coupler using a hammer. The impact from the first few strikes with the well driver firmly seated the pipe into the coupler for a permanent connection. An interference-fit coupler eliminated the need to screw pipe sections together, further eradicating any potential for stripped threads and poor fitment. The additional wall thickness of the coupler provided increased strength and stability to the well pipe assembly.

KEYWORDS: novel pipe coupler, potable water, sanitary water, shallow wells, tube well installation

1. INTRODUCTION

Nearly one-quarter of the world's population lacks reasonable access to the sanitary freshwater necessary for simple daily living (World Water Assessment Program, 2015). This is particularly distressing because a large majority of these people live in sub-Saharan Africa, where the water table is relatively high and easy to access, as shown in Figure 5.1 (IGRAC, 2017). It is well established that access to sanitary water can make a positive difference in the quality of life for families (UNICEF, 2021; World Health Organization, 2016). Subsurface groundwater is the logical choice for potable water for both human and animal consumption (Groundwater Foundation, 2019). Lack of access to groundwater is so fundamentally recognized as an ongoing human crisis that the United Nations, the National Academy of Engineering, and the Millennium Challenge Corporation have inspirational activities centered around improvement in the water supply (National Academy of Engineering, 2021; Millennium Challenge Corporation, 2017; World Water Assessment Program, 2015; World Health Organization, 2013).

Studies specific to Ghana show that the development of potable water sources contributes to both economic and quality of life benefits in communities by providing water, sanitation, and hygiene initiatives, which cannot happen without access to a plentiful supply of freshwater (Baldwin & Stwalley, 2022c; Baldwin & Stwalley, 2020). Local access to freshwater can also make a big difference in the agricultural production of a region, directly boosting the local economy (Baldwin & Stwalley, 2022b; Baldwin & Stwalley, 2021). In situations where shallow depths may hold freshwater, an installation known as a tube well may be used to access the aquifer (Wellowner, 2015; Koegel, 1985). Tube wells, also known as driven wells, have typically been installed

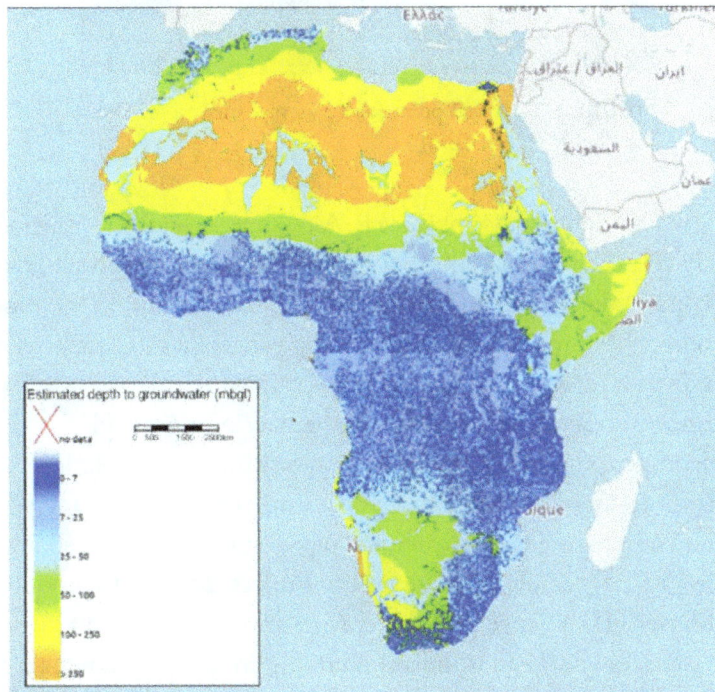

FIGURE 5.1. Estimated depth to groundwater (*mbgl*) and transboundary aquifer of Africa based on geologic and well drilling data (IGRAC, 2017). The groundwater is deepest in the Sahara and southwestern portions of the continent. It is shallower in sub-Saharan Africa..

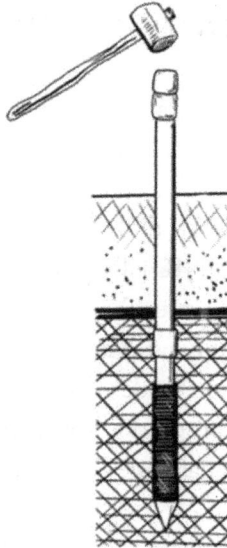

FIGURE 5.2. A cross-section view of a manual installation of a drive tube well installation (Proby, 2013)

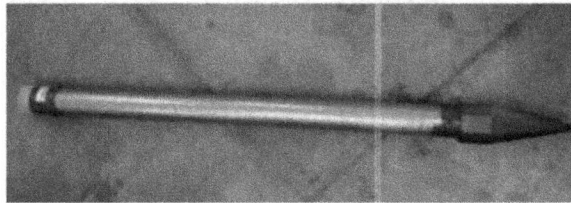

FIGURE 5.3. A well point for a 2 *in* pipe-driven tube well installation (Horn, 2019)

in the past by hand, as shown in Figure 5.2 (Proby, 2013). A team of several people can manually install a tube well by hand fairly quickly (US Army, 1994), but because of the strenuous effort involved, smaller numbers of people will tire quickly and take proportionally much longer to complete the process. Recent efforts by Purdue University researchers have mechanized the driving process and greatly improved the efficiency of tube well installation (Baldwin Kan-uge et al., 2023a), thereby making this water well process a possibility for alleviating the freshwater access crisis in certain areas of the developing world.

Tube wells are assembled from plumbing components on-site during installation. Driven tube well "stacks" typically consist of a well point, shown in Figure 5.3 (Horn, 2019); threaded galvanized pipe sections; and specialized drive couplers, shown in Figure 5.4, to fasten the pipe sections together (Baldwin & Stwalley, 2022a). The Purdue team modified a PUP vehicle (Purdue Utility Project, 2021) with an onboard hydraulics system to operate an enhanced Shaver HD-8 fence post driver (Shaver Manufacturing, 2009). The vehicle with the driver attachment in the transport position is illustrated in Figure 5.5 (Adams et al., 2019). Figure 5.6 shows the well driver erected into the operational position (Baldwin & Stwalley, 2022b). Unmodified from its original design for fence post installations, the channel piece is hydraulically lifted and then dropped onto the pipe stack. The channel is guided during its driving stroke by the framework of the overall implement. The impact of the channel from dropping under gravity and tensioned springs drives the pipe stack into the ground. The prediction of depth per hit using the well driver system has been established (Baldwin Kan-uge

FIGURE 5.4. A drive coupling for a 2 *in* driven tube well stack (Baldwin & Stwalley, 2022c)

FIGURE 5.5. Well-driver PUP vehicle in transport configuration with the hydraulic driving ram collapsed into holding cradle (Adams et al., 2019)

FIGURE 5.6. Well-driver PUP vehicle in operational configuration with the hydraulic driving ram erected (Baldwin & Stwalley, 2022b)

et al., 2023b). A downward movement of 2–5 *cm* per blow is typical within 2 or 3 blows from starting the driving process.

The forces associated with the mechanized process are significantly higher than those during hand installation. The driver with added weights had a static potential of 2400 *N* and an impact potential of 16,700 *N* (Baldwin Kan-uge, 2023). Although the specific driving process has continued to evolve with the team's experience, the vehicle integration with the post driver was successful (Horn & Stwalley, 2019). All field testing took place in West Central Indiana, with one well reaching over 7 *m* and successfully pumping 11.5 *L/min* (Horn, 2019). The ability of the team to hit water has steadily improved with experience, and the team has successfully produced and driven a single larger 4 *in* nominal pipe tube well (Baldwin Kan-uge, 2023). However, the 2 *in* pipe seems to be the optimal selection from an economic perspective, and the deeper the pipe was driven, the more apparent it was that the traditional drive couplers were inadequate for use with the mechanized process. The balance of this paper will describe the problems with the commercially available threaded drive coupler, the design and manufacture of the new novel prototype couplers, the preliminary testing of the design in a driven well installation, and conclusions about future development of the interference-fit tube well pipe coupler.

2. PROBLEM STATEMENT

Traditionally when making a tube well stack, each section of pipe is coupled together with the use of a well pipe coupler, shown in Figure 5.7. Because of the axial loadings that are imposed on these pieces, well pipe couplers are made differently than standard pipe couplers. Internally, the national pipe threads begin with a machined surface on the interior of the coupler. This is designed to be the mating surface for the connection of the coupler to the pipe rather than by carrying the load on the threads. However, the premanufactured galvanized pipes had very inconsistent thread length and taper, which made mating the sections of pipe in the specialty couplers problematic, and multiple issues in the design became relevant in well installations. If the pipe and mating surfaces were not in solid contact, then the force from the strike of the ram was transferred through the threads instead of the end of the pipe wall, eventually deforming the pipe material.

In an initial attempt to correct this lateral strength problem, specialty spacers were fabricated to carry the load between pipe sections from steel 1.5 *in* nominal galvanized conduit. Each coupler was configured in a lathe and bored out with a boring bar to allow the spacer to slide into the coupler. Because every pipe joint was different,

FIGURE 5.7. A commercially available 2 *in* nominal threaded driven well pipe coupler

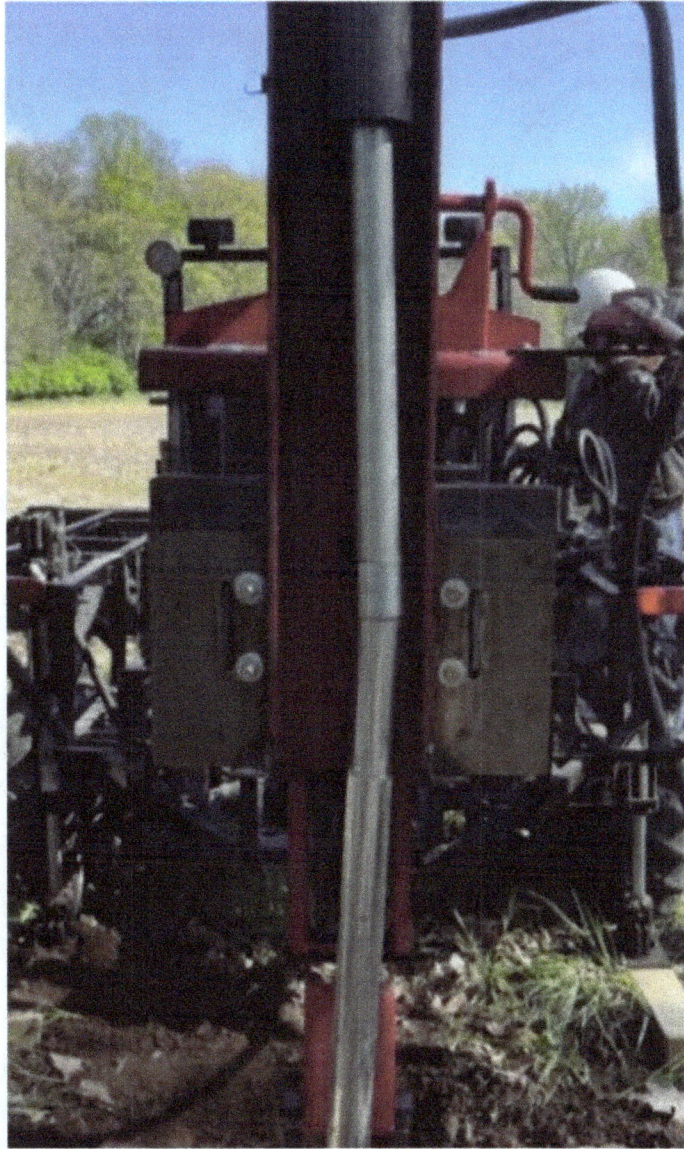

FIGURE 5.8. Stripped pipe threads during a well installation, resulting in a bent pipe stack

the spacers were cut to specific dimensions for each unique pipe joint. The increased complexity of pipe assembly during well installations using this system was challenging. If a pipe section became damaged or could not be used for whatever reason, the order and specifics of each spacer and pipe connection assembly became altered. This resulted in stripping pipe threads during well installations, when repeated blows would cause the coupler to shear the threads from the joining pipe section. The full strike force was transferred not through the edge of the pipe but instead through the national pipe threads. The damaged threads degraded the structural rigidity of the pipe stack, and bent pipes were the end result, as shown in Figure 5.8. The coupling between the pipe sections was truly a critical structural piece of the overall well pipe stack assembly. These difficulties made it apparent that a new and novel solution for the connection problem was needed. The balance of this paper details the team's proof-of-concept solution to the coupler issue and how well the novel design performed during an initial field test.

FIGURE 5.9. Interference-fit pipe coupler for 2 *in* nominal galvanized well pipe

3. DESIGN SOLUTION

Water wells are regulated in the United States by laws and codes from the state legislatures and state regulatory agencies in a reasonably consistent and uniform manner, accounting for local variations in conditions. Indiana codes were utilized for the construction of the well components in this study and their installation into production (Indiana Administrative Code, 2015; Indiana General Assembly Code, 2010, 1987; Indiana Code (Statutes), 1987), and one of the researchers holds a current Indiana Well Driller's License. All wells were installed using approved procedures and registered with the state, with the exception noted that the initial proof-of-concept coupler was not constructed from a food-safe material. The well owner at the proof-of-concept site was aware of this deficiency and intended only to use the tube well for watering a garden. The design criterion for the new coupler was that it must be capable of transmitting an impact force through it in a sufficient amount to facilitate the installation of a tube well stack through a mechanized process without deformation or losing axial alignment. The only constraints on the design were that it must be transferable to materials compatible with potable water design and disinfectants currently in use to initially sterilize new wells (American Ground Water Trust, 2012). For this design problem, a properly executed design will secure the pipe stack during installation, allow for ease of disinfection following the initial well surging, and not create a potential path for well contamination with aging. An efficient well pipe coupler will prevent broken stacks in midinstallation, which voids the work on-site to that point. A solid connection will prevent the drifting of the well point during

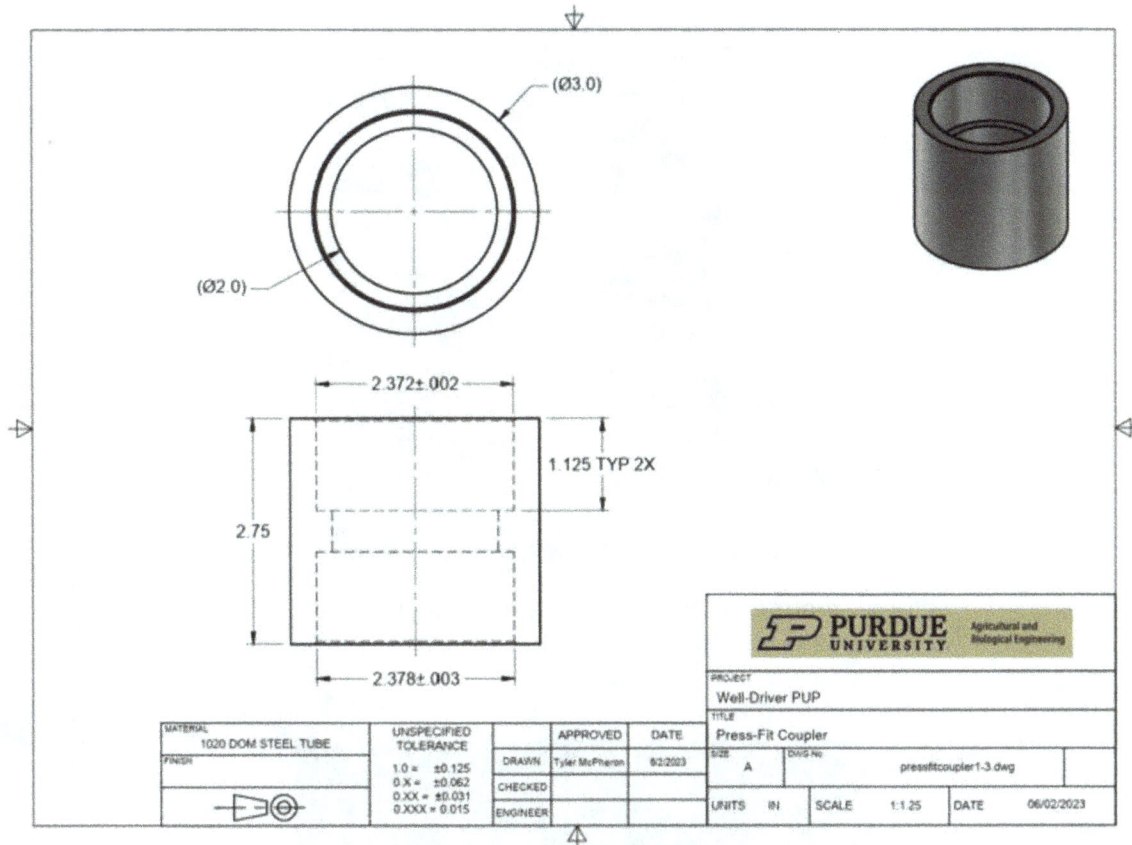

FIGURE 5.10. Prototype interference-fit tube well pipe coupler piece part drawing

FIGURE 5.11. Machining the prototype coupler with a boring bar on a manual metal lathe

FIGURE 5.12. Prototype tube well pipe coupler used to test fitment during machining

FIGURE 5.13. Prototype tube well pipe coupler with interference-fit pipe and slip-fit pipe installed

FIGURE 5.14 Interference-fit tube well pipe couplers welded to pipe sections, ready for installation into a pipe stack

driving, and it will remain connected once installed. In general, during the proper execution of the planned steps of an installation, the new well pipe couplers should save installers' time, materials, and sunk installation costs on abandoned wells due to broken part problems.

To achieve this positive solution, a prototype pipe coupler was machined out of carbon steel drawn-over mandrel tubing. The selection of the carbon steel as a material was expedient for this proof-of-concept trial and was not intended to be carried forward. The new coupler was designed with a very tight tolerance interference-fit on one end and a slip-fit on the other end. The interfering end was machined undersized 0.076 *mm* with a ±0.051 *mm* tolerance, while the slip-fit end was machined oversized up to 0.127 *mm*. The established tolerances for the 60 *mm* pipe and coupler qualify as a heavy force/shrink interference class fit (Krutz et al., 1994). This

FIGURE 5.15. Interference-fit pipe couplers during well installation, with a fully weighted driver

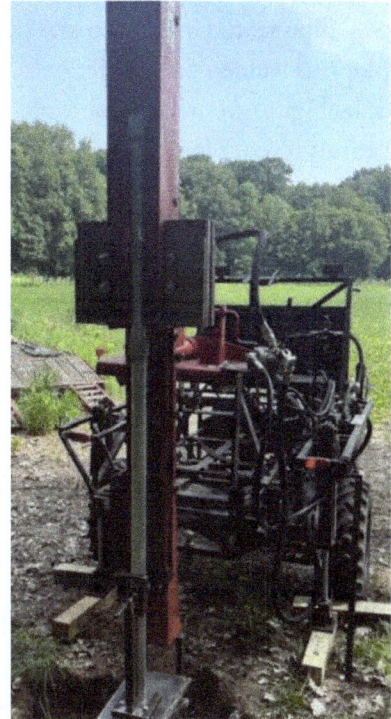

FIGURE 5.16. Prototype interference-fit pipe couplers during tube well installation, with multiple joints in stack

falls within the ISO 286 tolerance grade H7/P7 guidelines, and a surface finish for the mating surfaces was estimated to be 10–30 μm (Krutz et al., 1994). Each coupler was manually machined using a lathe and a boring bar. The well drive point and each pipe section were inserted in the slip-fit end and welded solid before going to the field for a well installation. The interference-fit end of the coupler remained exposed and available for coupling. At the well installation site, the joining pipe was inserted minimally into the undersized end of the coupler using a hammer. The impact from the first few strikes with the well driver seated the pipe into the coupler for a permanent connection. The driving forces created enough deformation on the clean pipe end within the fitting that the two pieces became effectively interference-fit held together. The technical drawing and a finished sample of this new prototype coupler in a short piece of pipe are shown in Figure 5.9.

The interference-fit pipe coupler prototype was designed to provide an alternative solution to the problematic use of the commercially available threaded pipe couplers during mechanized tube well installations, where the forces grew well beyond those seen in manual installations. Figure 5.10 shows the engineering drawing for the interference-fit pipe coupler and the critical dimensions used to manufacture couplers for use during the experimental well installations. The design was made for nominal 2 *in* pipe. Figure 5.11 shows the machining process for boring out the inner diameter using a lathe and a boring bar equipped with a carbide cutting bit. This operation was particularly sensitive, as the concept required an interference-fit coupler between the pipe and the couplers. Figure 5.12 shows the first prototype interference-fit coupler, which then was used for a dimensional quality assurance gauge for further manufacturing. This coupler was used to test-fit the newly made couplers and ensured that they were within the specified tolerances. Figure 5.13 shows the first prototype press-fit coupler with a short section of well pipe seated inside the interference-fit end and a short section of well pipe seated in the slip-fit end of the coupler. Once the slip-fit end was dimensionally approved, the coupler was welded onto a section of pipe. This end was therefore fully connected to the coupler and not removable. Figure 5.14 shows multiple finished 2 *in* nominal tube well pipe stock sections with the manufactured prototype couplers attached for use in field testing.

4. RESULTS

The prototype interference-fit tube well pipe coupler demonstrated its functionality in a 2 *in* driven well trial in June 2023. Figures 5.15 and 5.16 show the prototype pipe coupler in use during the well installation. This coupler design eliminated the need to thread the sections together, eliminating the potential for stripped threads and poor fitment from deformed material. The additional wall thickness of the coupler provided increased strength and stability to the well pipe stack at a previously weak junction. The limited experience reported here demonstrated a far superior resistance to slenderness flexure than the previously utilized threaded coupler. This proof-of-concept result would seem to verify the mechanical sufficiency of the prototype design. For a production unit, a stainless steel or galvanized piece would ultimately be required. The seal created by the deformation of the pipe in the mated coupler under repeated blows from the driver was solid and felt integrated through the joint under attempted twisting. An actual series of pressure checks would be required to confirm this, but based on its resistance to motion, the joint seemed likely to be leak-free and liquid-tight. Finally, the installation drive process using the interference-fit couplers was smooth and quicker. It did not have any of the minor interruptions or difficulties that had plagued earlier installations. If this operational simplicity remains through further testing, then the real savings for this innovation would be from the overall improvement in the well driving process execution. Additionally, it is fair to state that when less goes wrong inside a process, it is typically a safer process for those performing it.

5. CONCLUSION

Since this was only a single proof-of-concept execution of an idea for a novel well pipe coupler, the fairest thing that can be said is that this prototype concept looks promising enough to pursue further investigation. In order to perform a series of tests needed to statistically demonstrate the concept, improved manufacturing processes would need to be developed for making these novel couplers. These units were manually turned on a machine lathe. Computer numerical control programming and manufacturing would need to be developed to make the volume of parts needed to thoroughly test the concept. Additionally, as a proof-of-concept execution, these units were made from low-carbon steel. This material would need to be switched to a stainless steel or postmachining galvanized part to be considered safe for potable water use. Overall, the preliminary testing of this interference-fit tube well pipe coupler was successful. The main conclusions from the effort can be summarized as follows:

- The interference-fit coupler reduced the number of field joints in a tube well stack by half.
- The interference-fit coupler eliminated the pipe threading operation in tube well stack preparation.
- The interference-fit coupler created a stiffer well pipe stack than those with specially designed threaded pipe couplers.

- The couplings became fully solid after multiple impacts from the mechanized tube well driver.
- The interference-fit tolerances in the coupler are challenging to hold during manufacture.
- Utilization of the interference-fit tube well pipe coupler made the overall installation process quicker, smoother, and able to proceed in a trouble-free manner.

Although developed to solve a unique problem associated with the mechanized installation of tube wells, this innovation could have application in the realm of drilled wells. Steel casings, which are installed for the initial depths of mechanically bored wells, are still dependent on threaded couplings. These assemblies do not see the forces found in driving a pipe stack, but the simpler installation might have significant time-based advantages on a well drilling job site. For either market, a commercially produced product would require, at minimum, a more extensive statistically valid testing program with a preproduction version of the device. Production difficulties from the tight tolerances on the interference-fit coupler could be eliminated or reduced through advanced machining techniques. With proper promotion, these devices might find significant enough of a market between those two uses to become a successful niche market commercial plumbing product.

6. ACKNOWLEDGMENTS

This work was supported in part by the Purdue University Office of Engagement. The Purdue Agricultural & Biological Engineering Department has also graciously helped provide backing for this project. Additionally, thanks are due Dr. John Lumkes for kindly providing a PUP vehicle to the project. Dr. Carol S. Stwalley is acknowledged for her professional assistance in editing and formatting this manuscript. Discussions with the professional staff of the Water Division of the Indiana Department of Environmental Management have been insightful and productive in the development of this innovation. An artificial intelligence–powered large language model software, ChatGPT, by Microsoft, Incorporated (Redmond, Washington) was used to aid in the organization of this effort, but all manuscript elements are original from the authors. Any errors are entirely the result of the authors. The mention of trade names or commercial products in this article is solely for the purpose of providing specific technical information and does not imply recommendation or endorsement by Purdue University. The findings and conclusions in this publication are those of the authors, and they should not be construed to represent any official Purdue University determination or policy. Purdue University is an equal opportunity/equal access organization.

7. NOVEL TUBE WELL PIPE COUPLER QUESTIONS

1. Explain the difference between standard pipe couplers and traditional tube well pipe drive couplers.
2. If a tube well pipe stack becomes out of vertical alignment by 5°, what percent of the impact force is transferred horizontally?
3. How does the slenderness ratio of the tube well pipe stack affect the driving results, particularly during the addition of a new pipe section to the stack?
4. What expediency was taken during the proof-of-concept testing of the novel coupler that would prevent the initial wells from being certified as sanitary? How could that problem be overcome for production water wells?

5. Describe five advantages of the new pipe coupler design and one disadvantage. How might that problem be overcome for production installation of tube wells?

8. REFERENCES

Adams, J., Hampston, T., & Wilson, H. (2019). *Improvements to the Purdue Utility Platform well-driver.* ABE 485 Capstone Experience Final Report, Purdue University, Agricultural & Biological Engineering, West Lafayette.

American Ground Water Trust. (2012). agwt.org. Accessed December 15, 2021, from Water Well Disinfection Procedure. https://www.agwt.org/content/water-well-disinfection-procedure.

Baldwin Kan-uge, G. L. (2023). *Improvements to the driving capabilities of a well-driver PUP (Purdue Utility Project) to install low-cost driven water wells* [PhD dissertation]. Purdue University.

Baldwin Kan-uge, G. L., McPheron, T. J., Horn, Z. J., & Stwalley, R. M., III. (2023a). Cost-effective, sanitary shallow water wells for agriculture and small communities using mechanized tube well installation. In J. Tarhouni, *Groundwater: New advances and challenges.* Intech Open. https://doi:10.5772/intechopen.109576.

Baldwin Kan-uge, G. L., McPheron, T. J., & Stwalley, R. M., III. (2023b). *Design and development of a preliminary regression model to determine the driving capabilities of the well-driver PUP* [Paper presentation]. ASABE Annual International Meeting, Omaha. ASABE. https://doi:10.13031/ aim.202300014.

Baldwin, G. L., & Stwalley, R. M., III. (2020). *Promoting restoration of Lake Bosomtwe through spatial analysis of existing water, sanitation, and hygiene (WASH) sources in Ghana, West Africa* [Paper presentation]. ASABE 2020 AIM, Pasadena. https://doi:10.13031/aim. 202000589.

Baldwin, G. L., & Stwalley, R. M., III. (2021). *An economic analysis: The scale-up of irrigation systems in Ghana, West Africa* [Paper presentation]. ASABE Annual International Meeting, Pasadena. https://doi:10.13031/aim.212100009.

Baldwin, G. L., & Stwalley, R. M., III. (2022a). *Improving the driving capabilities of a well-driver PUP (Purdue Utility Project) to install low-cost driven water wells* [Paper presentation]. ASABE Annual International Meeting, Houston. https://doi:10.13031/aim.202200203.

Baldwin, G. L., & Stwalley, R. M., III. (2022b). Opportunities for the scale-up of irrigation systems in Ghana, West Africa. *Sustainability, 14,* 8716. https://doi:10.3390/su14148716.

Baldwin, G. L., & Stwalley, R. M., III. (2022c). A review of freshwater programming and access options in Ghana, West Africa. *African Journal of Water Conservation and Sustainability, 10*(4), 1–19.

Groundwater Foundation. (2019). https://www.groundwater.org. Last accessed from get-informed/basics/groundwater.html.

Horn, Z. J. (2019). *Prototyping a well-driver PUP (Purdue Utility Project) to install low-cost driven water wells* [Master's thesis]. Purdue University. https://doi:10.25394/PGS.8038991.v1.

Horn, Z. J., & Stwalley, R. M., III. (2019). *A low-cost mechanized tube well installer* [Paper presentation]. ASABE 2019 AIM, Boston. https://doi:10.13031/aim.2019011319.

IGRAC. (2017). *Africa groundwater portal.* Last accessed December 15, 2021, from International Groundwater Resources Assessment Centre of the United Nations, https://www.un-igrac.org/special-project/africa-groundwater-portal.

Indiana Administrative Code. (2015). Article 13. *Water well drills and water well pump installers.* Last accessed December 15, 2021, from https://www.in.gov/health/eph/files/312-IAC-13.pdf.

Indiana Code (Statutes). (1987). *Article 39. Water well drilling and pump installer contractors.*

Indiana General Assembly Code. (2010, 1987). *Title 25. Article 39: Water well drilling contractors.* Last accessed December 15, 2021, from https://www.iga.in.gov/legislative/laws/2020/ic/titles/025#25–39.

Koegel, R. G. (1985). *Small diameter wells.* Last Accessed December 15, 2021, from Food and Agriculture Organization of the United Nations (FAO), https://www.fao.org/3/X5567E/x5567e00.htm#Contents.

Krutz, G. W., Schueller, J. K., & Claar II, P. W. (1994). *Machine design for mobile and industrial applications.* Society of Automotive Engineers, Inc.

Millennium Challenge Corporation. (2017). *Measuring results of the Ghana water and santiation sub-activity*. Last accessed November 17, 2019, from mcc.gov/resources/doc/summary-measuring-results-ghana-water-sanitation-sub-activity: https://www.mcc.gov/resources/pub-pdf/report-ghana-closed-compact.

National Academy of Engineering. (2021). *NAE grand challenges for engineering*. Last accessed February 12, 2021, from http://www.engineeringchallenges.org/challenges.aspx.

Proby, F. (2013). *Shallow well drilling*. Last accessed October 11, 2021, from lifewater.org: https://www.lifewater.org/wp-content/uploads/2018/10/lifewater-Well-Drilling-manual.pdf.

Purdue Utility Project. (2021). *Where*. Last accessed April 30, 2022, from https://engineering.purdue.edu/pup/where/.

Shaver Manufacturing. (2009). *Operator's manual for hydraulic post driver model HD-8 & HD-8-H*. Last accessed December 15, 2020, from http://www.shavermfg.com/media/uploads/HD8-Operator-Manual.pdf.

UNICEF. (2021). *Collecting water is often a colossal waste of time for women and girls*. Last accessed December 16, 2021, https://www.unicef.org/press-releases/unicef-colelcting-water-often-colossal-waste-time-women-and-girls.

US Army. (1994). *US Army field manual: Well drilling operations (FM5–484)*. US Department of Defense.

Wellowner. (2015). *Types of wells*. Last accessed November 17, 2019, from wellowner.org/basics/types-of-wells.

World Health Organization. (2013). *UNICEF progress on sanitation and drinking water—2013 update*. Last accessed December 15, 2020, from https://www.data.unicef.org/resources/progress-on-sanitation-and-drinking-water-2013-update/.

World Health Organization. (2016). *What is the minimum quantity of water needed?* Last accessed January 1, 2019, from https://www.who.int/water_sanitation_health/emergencies/qa/emergencies_qa5/en/.

World Water Assessment Program. (2015). *The United Nations world water development report 2015: Water for a sustainable world, facts and figures*. Last accessed December 15, 2020, from https://www.unwater.org/publications/un-world-water-development-report-2015.

Proof-of-Concept Testing for a Novel Well Pipe Coupler Used in Shallow Tube Wells

Tyler J. McPheron, Grace L. Baldwin Kan-uge, and Robert M. Stwalley III

PURDUE UNIVERSITY®
Agricultural and Biological Engineering

Objective

Improve the Well Pipe Stack Integrity for More Efficient Mechanized Installation

Tube Wells

Tube Wells are also known as Driven Wells. Instead of boring into the earth, a tube well point pushes the earth aside as it penetrates downward into an aquifer. Mechanization increases the productivity of the installation by raising impact forces on the pipe stack. This can cause a stripping of threads in traditional impact couplings, if the stack moves slightly out-of-alignment. Demonstration wells using the Well Driver PUP indicated that a stronger coupler driving was required to hold the slender stack straight under heavy impact forces.

African Tube Well Potential

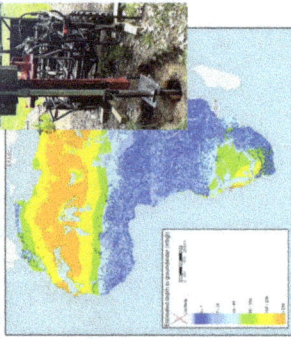

United Nations data demonstrate that nearly 1 in 6 people worldwide lack easy-to-access sanitary water. One third of those people are standing within 15 m of groundwater and lack the means to access it.

Interference-Fit Drive Coupling

The interference-fit coupler is a machined pipe section, with a tight-fit side, welded into place, and an interference-fit side, driven into place. This machined piece added sufficient stiffness to the well pipe stack to prevent buckling out of line during the driving process.

Image of Coupling Piece

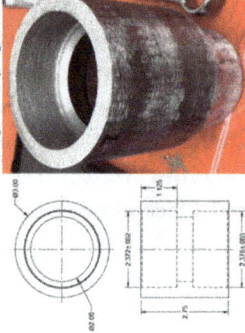

Image of Coupling Piece with Mated Pipe

Well Driver PUP

The pre-made 1.5 m sections of well pipe were the optimal length for the Well-Driver PUP machine. A different unit might require a different drive pipe length. Although the production of the pipe assemblies was slow (> 2 h / ea.), the driving speed was nearly doubled, and the well installation process time was cut by one quarter.

Conclusions

The prototype well pipe coupler reduced the number of field fittings by two. It eliminated the threaded couplings. It created a stiffer well pipe stack. The coupling became solid following the deformation created by the impact of the driving ram. The entire process became less troublesome, faster, and smoother. More development is needed to reduce the pipe assembly production time.

Dr. John Lumkes of the Purdue University Agricultural and Biological Engineering department is thanked for his gracious contribution of the Purdue Utility Platform (PUP) vehicle.

This project has received support from the Purdue University Office of Engagement and the Agricultural and Biological Engineering department.

CHAPTER 6. SHALLOW WELL TECHNOLOGY

This chapter addresses the larger global situation regarding freshwater access across the developing world and reveals both the linkage between water use and distance to the water source and how the burden for daily water collection falls on the most disadvantaged members of households. This chapter reviews the various means of providing groundwater for use and discusses the various advantages and disadvantages of each technique. The technology of tube (sand) wells is explored, and the Purdue-developed mechanized tube well installer device is reviewed. The initial outcomes from the tube well development effort are presented, including the water quality results, where tube well installation might be feasible, and an economic analysis of tube well installation. Speculation about the potential impact on developing communities and society from the widespread adoption of technology is provided. This review was initially published as a chapter in *Groundwater—New Advances and Challenges* and provides a general reference for water well technology. The review includes information on identifying specific challenges in providing reasonable access to clean freshwater for households in the developing world as well as a detailed overview of the Purdue mechanized tube well installation technology.

COST-EFFECTIVE SANITARY SHALLOW WATER WELLS FOR AGRICULTURE AND SMALL COMMUNITIES USING MECHANIZED TUBE WELL INSTALLATION

GRACE L. BALDWIN KAN-UGE, TYLER J. MCPHERON (PURDUE UNIVERSITY AGRICULTURAL & BIOLOGICAL ENGINEERING), ZACKARIAH J. HORN (PURDUE UNIVERSITY AGRICULTURAL & BIOLOGICAL ENGINEERING), AND ROBERT M. STWALLEY III

ABSTRACT

Multiple studies have adequately demonstrated the connection between sanitary water supply for developing communities and sustainable economic growth. Unfortunately, the cost of traditional drilled water wells prevents their more rapid installation across much of the developing world. Numerous communities and agricultural areas could benefit greatly from access to groundwater less than 10 *m* deep. Researchers have developed a means to mechanize shallow tube well installation to provide sanitary water wells of modest capacity. A hydraulic ram for agricultural fence post driving has been attached to a small Purdue Utility Platform (PUP) vehicle and repurposed to drive a small-diameter well pipe. This chapter will outline the water access problem from a global perspective, describe the traditional means of construction for sanitary water wells in remote areas and their relative costs, and detail the recent advancements and potential cost savings provided by a simple mechanized means to install tube wells in shallow water table areas.

KEYWORDS: irrigation, potable water, sanitary water, shallow wells, tube well installation

1. INTRODUCTION

Globally, 1.8 billion people (22.5%) use an unimproved source of drinking water with no protection against contamination from feces. Safe drinking water combined with good hygiene and improved general sanitation is generally known as water access, sanitation, and hygiene (WASH). Improved WASH conditions could potentially prevent around 842,000 deaths each year [1]. The WASH acronym specifically stands for safe water access for drinking and household use that is free from chemical and biological pollutants, sanitation including access to a toilet (latrine) that safely separates human excreta from the environmental, and hygiene focusing on public health and prevention of the transmission of fecal-oral diseases [2]. This chapter will examine the state of the water component of WASH programming in the developing world. Traditional techniques to access groundwater will be reviewed, and recent work using mechanized tube well installation will introduce the well-driver PUP technology [3]. Implementation of this technology could provide meaningful progress toward addressing the sixth United Nations sustainable development goal: "To ensure availability and sustainable management of water and sanitation for all," which incidentally will help drive progress across many other sustainable development goals [1]. People must have equitable and affordable access to safe and sufficient water that is palatable and in sufficient quantity for both drinking and domestic purposes [4]. Ongoing research proposes to increase access to subsurface water by improving the operational capabilities of the well-driver PUP vehicle [5]. Implementation and dissemination of this novel vehicle technology could improve access to safe water for drinking and domestic purposes in developing countries and can play a key role in WASH programming.

1.1. WATER QUALITY ACCESS

In developing countries, access to safe water is critical to the quality of life and the potential for economic growth. Water-related diseases pose a major risk to individuals in developing countries due to the consumption and use of unsafe and poor-quality water sources [4]. Often, water sources are prone to contamination due to the movement of contaminates through surface transport processes. Water contamination due to surface runoff, leaching, and pollution from agrochemicals into groundwater sources can lead to increased risks of humans contracting waterborne pathogens from drinking water. Poor waste management and the inappropriate disposal of human and animal excreta can result in higher levels of contamination in water resources. The presence of excreta in water used for human consumption often leads to serious but preventable diseases such as typhoid and cholera. Water that is high in fecal coliform bacteria, which is generally greater than 99% *Escherichia coli*, indicates a level of human and animal waste contamination in the water and the possible presence of other harmful pathogens [4].

Excess fertilizer use can lead to the leaching of dissolved nitrogen through the soil profile, resulting in additions of nitrate into groundwater resources. The consumption of drinking water containing nitrate higher than 2 *mg/L* for adults has also been shown to lead to adverse health effects, specifically higher risks of cancers [6]. For mammals, the adverse health pathway is nitrate in drinking water increasing the production of N-nitroso compounds, which are highly carcinogenic [7]. Infants ingesting drinking water containing a high nitrate content can have low oxygen levels in their blood, leading to a potentially fatal condition known as "blue baby syndrome." Access to water that is safe and considered of good quality is essential to overall community health and healthy living conditions for people and domestic livestock around the world.

1.2. WATER QUANTITY ACCESS

In many locations around the world, people use unsafe water sources or lack sufficient access to water for both drinking and domestic purposes, creating very unhealthy circumstances for these individuals. This is because in developing countries, clean water access is not always possible. Water resource use is often constrained due to the terrain and hydrology of a specific location. Particularly in sub-Saharan Africa, women and children often walk great distances to obtain access to water for household use. In many regions, water is carried on top of one's head while simultaneously leading and watering livestock. According to the Sphere humanitarian

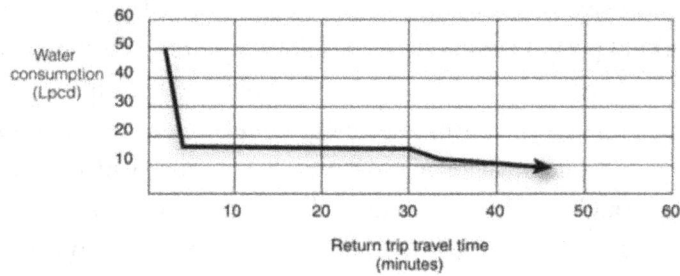

FIGURE 6.1. The relationship between water collected, journey time, and domestic consumption [8]

TABLE 6.1. *Simplified table of basic survival water needs [4]*

Survival needs: Water intake (drinking and food)	2.5–3 *L* per day	Depends on the climate and individual physiology
Basic hygiene practices	2–6 *L* per day	Depends on social and cultural norms
Basic cooking needs	3–6 *L* per day	Depends on food type and social as well as cultural norms
Total basic water needs	**7.5–15 *L* per day**	

standards, for any water-based source, the distance from the household to the nearest waterpoint should not exceed 500 *m*, and the queue time for water sources should be no greater than 30 *min* [4]. In situations requiring longer travel times, individuals are far less likely to collect larger amounts of water, as seen in Figure 6.1.

These data indicate that families, especially those living farther away from a water source, will only collect the basic minimum amounts of water required for survival. Within a just society, all people would have an equitable and affordable means of access to a sufficient supply of water that could be used for drinking, hygiene, and domestic purposes [4]. Table 6.1 displays the Sphere recommended minimum total water need for basic survival. The average water used for drinking, cooking, and personal hygiene in any household is 15 *L* per person per day and for an average month of 30 *days* is 450 *L*. For a family of five, 2250 *L*, or 2.25 *tons*, of water would be required to meet the minimum demand for all domestic uses. Obviously, the amount of water required for an individual can vary based on the community and the context, but human living needs require a minimum level of water for survival [4]. When this water is not located in the home and must be collected elsewhere, productive time for alternative activity is lost [9]. Women and children are disproportionately impacted by this cruel labor requirement. For women, it shortens the available time for them to be with their families, provide childcare, perform household activities, and engage in entrepreneurial enterprises. Water collection by both boys and girls can take time away from their education, and sometimes can even prevent them from attending school altogether. The brutality of having to collect water and transport it on a daily basis robs both women and children of their most valuable resource [9].

1.3. CURRENT METHODS OF SOURCING GROUNDWATER

There are three primary types of wells utilized to obtain groundwater resources for both drinking and domestic purposes. These methods include hand-dug wells, drilled wells, and driven tube wells. Hand-dug wells are constructed manually and require individuals to dig until below the water table [10]. These wells must be dug during a dry season to ensure that the water table is at the lowest possible level. In situations where the water table has receded below the depth of the well, the bottom of the well must be dug deeper to access the water table [11]. Hand-dug wells have a circular cross-section and should be lined with stone, brick, or tile to prevent the earthen sidewalls from collapsing inward. This type of well does not have a continuous casing and grouting, making it far more prone to contamination from surrounding surface sources [12].

The most common method for obtaining groundwater is by creating a drilled well. Unfortunately, this technique is always the most expensive of the methods considered, but you get what you pay for. This type of installation produces a dependable, long-term, sanitary well and is considered the gold standard of groundwater access. Through this means, groundwater is accessible to deeper levels than by other options. Throughout the world, the general preference for WASH programming is to install a deep-drilled well to access groundwater having a lower likelihood of containments. However, in developing countries there are often not enough reputable companies with available drilling equipment to meet the demand for installed wells at an affordable cost in a timely manner. In Haiti, for example, there are very few drilling companies. Even when an organization or individual has sufficient funds to install a deep well, the wait time for a drilling company to come and install a new well can be over one year [13].

A driven tube well can be an acceptable alternative to traditional drilled wells under certain conditions [14]. Tube wells are constructed with a well point connected to galvanized steel pipe, which serves as the well casing. The well point is sharpened and driven through the soil, accessing groundwater through a fine mesh or perforation on its circumference near the point. This type of well is commonly driven by hand, with a tripod system setup, and installation tends to be a very labor-intensive effort. Reducing the labor element in the tube well installation process could potentially make this type of well more feasible across a considerable range of developing territory [15, 16]. The Purdue well-driver PUP mechanizes the tube well installation process by using a hydraulic ram. This mitigates the intensive labor component generally accompanied with driven wells and dramatically reduces the installed cost of sanitary water sources through the improved productivity of equipment and personnel involved. The remainder of this chapter describes the current status of groundwater access, the well-driver PUP technology, and the economic potential of the technology.

1.4. INDIVIDUAL ACCESS TO GROUNDWATER FROM WELLS

The physical water access point is a critical element of all wells, but it is particularly vital to community wells and those with shared access. Modern well standards require that the designer do everything possible to prevent contamination on the surface from entering the well aquifer. This criterion alone discourages the investment of effort and resources into hand-dug wells, as it is far more difficult to maintain sanitary access conditions for these types of wells. For all wells in general, surface-to-aquifer contamination results from two sources: down-the-borehole backwash and beside-the-casing downward drainage. To seal the casing from the surface, all modern wells should have adequate concrete pads surrounding the casing as it rises above the surface of the ground. The pads need to be extra strong in community well situations to withstand the burden of heavy traffic near the wellhead. The pads need to be properly sloped so the drainage is carried away from the wellhead and does not accumulate nearby. The concrete pads for pumps must be left to cure adequately before the pump head and water outlet assemblies are installed. Down-the-borehole contamination is best prevented by using a check valve, sometimes called a backflow preventer, in the pump assembly. The process that must be prevented is a syphon from an aboveground water storage tank back down into the aquifer. Any contamination present in a storage tank could potentially be injected into an underground aquifer during a syphon event at a wellhead. A community water bucket dropped into a hand-dug well poses essentially the same risk of contamination. For a drilled or driven well, a hand pump or an electric pump in the casing is the recommended means to keep a water well access draw-point sanitary and safe for all patrons.

2. REVIEW OF GROUNDWATER ACCESS TECHNOLOGIES

This section will contain a review of groundwater access options and the types of wells and drilling methods used throughout the world. Hand-dug wells and drilled wells will be explored. The components required to install a tube well are introduced along with a discussion of previous tube well installations using the prototype well-driver PUP.

2.1. GROUNDWATER-ACCESSED DRINKING WATER OPTIONS

Water sources are often classified as "improved" or "unimproved." Improved sources are piped public water into homes, public standpipes, water wells or boreholes, protected (lined) dug or hand-dug wells, protected springs, bottled water, and rainwater collection [17]. Unimproved sources are unprotected wells or springs, sachet water, vendors, tank trucks, and surface waters [17]. International WASH efforts tend to push communities toward the installation of improved water sources. Common components in current WASH programming efforts include community involvement through the establishment of community-led WASH committees, the construction of new water access points, the rehabilitation of preexisting water sources, the installations of new wells or community boreholes, small town water systems, and pipe extensions [18].

In many cases, women, poor households, and marginalized groups disproportionately experience the negative impacts of inadequate WASH resources. This primarily occurs because these groups are more than likely to have limited access to WASH services [19, 20, 21, 22]. Marginalized groups often have less input at both the household and community levels and in decision-making processes and the governance of resources relating to WASH [23]. Studies show that income, education, household size, and region are all significant predictors of access to improved water and sanitation [24, 25]. Therefore, many WASH programs and interventions utilize the methodology of empowering beneficiaries, which increases equitable access and the sustainability of water and sanitation infrastructure solutions [26, 27, 28].

2.2. GROUNDWATER

Water that is below the water table in soil is generally called "groundwater" [29]. This is underground water that can be removed by wells. The groundwater zone acts as a natural reservoir or system filled with fresh water. An aquifer is a saturated bed, formation, or group of formations which yields water in sufficient quantity to be used for economic purposes [14, 29]. Water storing formations and groundwater reservoirs are synonymous for "aquifer." There are two main types of aquifers: confined and unconfined [29]. An unconfined aquifer is where water enters from the soil surface and passes through the soil profile to enter the aquifer. A confined aquifer has an impermeable geological layer that prevents surface water from directly flowing into the aquifer. The installation of a well or borehole under these conditions includes drilling through the geological layer confining the aquifer in order to move the water from the deep aquifer up to some higher level. In this way, water wells are accessed for groundwater across the globe for both drinking and domestic water uses. Properly accessed groundwater is sanitary and is generally sustainable.

2.3. TYPES OF WELLS

In many countries particularly those in sub-Saharan Africa, individuals obtain their drinking water from community wells, which include both protected wells and boreholes. These water access points are most commonly located outside of dwellings and are in the form of a public tap or standpipe. The terms "wells" and "boreholes" tend to be used interchangeably worldwide. A borehole is the generalized term for any narrow shaft drilled into the ground. It generally contains both a pipe casing and a well screen to prevent the entry of soil into the water flow [30]. There are three primary methods of well construction: dug, drilled, and driven wells. Water wells can be installed either through manual methods or with powered tools [11].

2.3.1. DUG OR HAND-DUG WELLS

Traditionally, dug wells are excavated by hand using simple tools such as a pick and shovel, with a bucket on a rope to remove cuttings [11]. Figure 6.2 portrays an example of a hand-dug well installation. Although some pieces of dug well construction may be mechanized to a certain degree, the process to construct this kind of well is very intensive manual labor. A dug well is excavated below the water table during the dry seasons until the incoming water exceeds the digger's bailing rate. These wells should be circular in cross-section and lined with stones, bricks, tile, or other material to prevent the well from collapsing inward. This type of well does not have a continuous casing and grouting, making it more prone to contamination from surrounding surface sources. Dug wells have larger diameters and expose larger areas of the aquifer to the excavation. Therefore, these wells are able to obtain water from less permeable materials, such as very fine sand, silt, and clay [12].

FIGURE 6.2. Cross-section view of hand-digging a water well [11]

Most wells of this type are shallow and not able to achieve the depths that a bored or driven well can. This type of well often goes dry during droughty seasons because the water table drops below the well bottom. It is during this type of period that maintenance on the well can be performed. However, working on dug wells is quite risky. Someone must be lowered into the well to work. Labor under these installations is potentially dangerous due to the high potential for cave-ins and the lack of oxygen. Since it is difficult to dig very deep, hand-dug wells generally extend no farther than 30 m in depth [11].

To access more water in situations where the water table has dropped lower than the depth of the well, the bottom of the well must be excavated deeper to reach the new aquifer level. Water is typically lifted to the surface by attaching a bucket to a rope and drawing the water up by hand or crank. Unfortunately, obtaining water in this manner can also transmit bacteria into the groundwater source. Contamination of the water source is best prevented by sealing the walls, pouring a concrete apron around the base of the well, providing a raised parapet above the face around the well, using a lid over the top of the well, and utilizing a hand or electric pump to obtain water. Obviously, these features add additional costs to the well [11].

2.3.2. DRILLED WELLS

A well drilling machine is normally referred to as a "drill rig" or just a "rig" [11]. Drilled wells are able to penetrate consolidated material and require the installation of casing and a screen to prevent the inflow of sediment and to keep the well from collapsing inward [12]. This type of well can be pushed to more than 300 m in depth. The surface area around the casing has a segmented or concrete pad that is constructed to prevent contamination by water draining from the surrounding surface downward around the outer portion of the casing. The pad is most often constructed from neat cement or bentonite clay [12]. Pads are typically left to cure for a period of time prior to well commissioning, during which a well casing cap remains on the newly drilled well to prevent contamination. After the pad has cured, the pump cap can be removed and a pump head can be installed. Installing a pump head too soon, prior to pad curing, can lead to breakage of the concrete pad in use. Thus, it is vital to provide a proper cure time for the concrete when installing pump equipment and subjecting the pad to heavy operational loadings. Powered well drilling methods include the percussion cable tool, jetting, mud rotary, and air rotary techniques [11]. The most common powered installation methods used today are the percussion cable method and the mud and air rotary methods, but drilled wells can also be installed by hand [12, 31, 32].

2.3.3. BORED OR HAND-AUGERED WELLS

This method of drilling a well uses a small-diameter open-bottom bucket with angled teeth to manually cut into the soil. An example of this type of installation is shown in Figure 6.3. The bucket is attached to a T-shaped handle at the top through a series of steel rods, which can be rotated manually and pushed downward. As the bucket fills, the contents are lifted out and emptied. Additional rods are added as the hole deepens. This method is sometimes used for soil sampling in addition to shallow well construction. The diameter of the hole produced by this method is typically less than 8 cm, and the process is very depth-limited, as the drilling rate and material removal are very slow. Once below the water table, it is generally problematic to go deeper, and it is difficult to prevent the hole from collapsing inward. Also, the soil profile and composition greatly affect the depth of a well that can be installed using this method. In loose silt or sand, it is possible to go up to 10 m in depth, but in more compacted soil it would be quite difficult to reach this depth with manual drilling [11]. Therefore, most drilled well installations have been adapted to utilize machinery instead of human power.

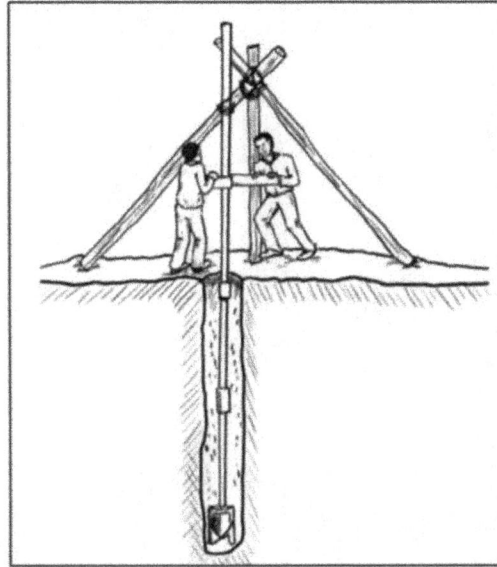

FIGURE 6.3. Cross-section view of a hang-augered water well installation [11]

2.3.4. PERCUSSION (CABLE METHOD)

This well drilling method utilizes repeated lifts and drops of a chisel-edged bit to break loose and pulverize material in the bottom of the hole. A small amount of water is added to the hole to form a slurry of the excavated material. The percussion bit is removed periodically, and a bailer is lowered into the hole to remove the slurry mixture containing the excavated material. The excavated material is brought to the surface and discarded. Bailing is repeated until the hole has been thoroughly cleaned. Bailing and drilling are alternated in this fashion until the desired depth is reached. If the hole is unstable, a casing can be lowered into the hole to prevent it from collapsing. The percussion drilling method is able to penetrate all types of materials, but in very hard stone progress can be quite slow. The percussion technique is frequently associated with a large truck-mounted attachment or motorized trailers similar to that shown in Figure 6.4 [31]. "The [percussion] machinery ranges from a basic skid-mounted powered winch with a tripod, to a complex set of pulleys and runs with a large mast" [11]. These larger cable tool rigs have hydraulic motors to raise and lower the mast and rotate the drums of the cable. Fewer cable tool rigs are being utilized in developed areas of the world today because compared to hydraulic rotary drill rigs of similar size, percussion drill rigs work slower [11]. Additionally, when drilling in loose sediments, it is necessary to drive a steel pipe behind the drill bit to prevent the borehole from collapsing. The sections of this "drive casing" must be welded together going in and cut apart coming out, which requires that an arc welding and cutting torch set be available during the drilling process [11]. These additional processes are not required when using alternative well drilling technologies.

2.3.5. JETTING

This well drilling technique utilizes a high-pressure pump to force water down a drill pipe and out a small diameter nozzle in order to make a "jet" of water that loosens the soil. The return flow of water outside the drill pipe carries the cuttings up to the surface and into a settling pit. A circulation pump returns the water back down the pipe to continue bringing more cuttings to the surface. A tripod setup is typically used and rotated by hand to ensure a straight borehole. In addition to the water piping components, this well drilling technique

FIGURE 6.4. Truck-mounted cable tool rig (left) and trailer-mounted tool rig (right) [31]

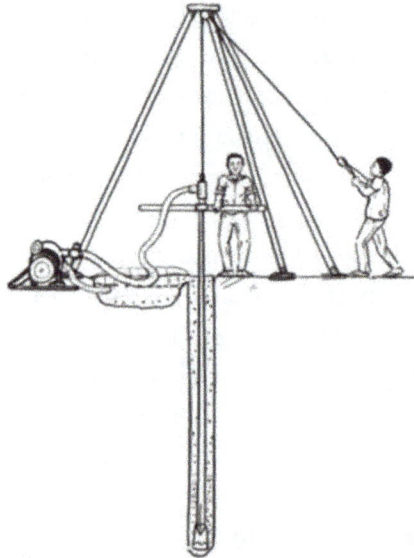

FIGURE 6.5. Cross-section view of jetting a drilled water well [11]

requires a high-pressure water pump and two people to set up and operate the rig. Figure 6.5 illustrates this method of well drilling. Unfortunately, this method is only suitable for fine-grained and soft sediment soils and requires a nearby water source to supply the jet system. This well drilling technique is not suitable for gravel and in hard soil profiles [11].

FIGURE 6.6. An on-site isometric view of a mud rotary drilling rig in an operational configuration [11]

2.3.6. MUD ROTARY

A mud rotary drilling rig includes a "jet" in combination with a larger-diameter cutting bit, precut and threaded lengths of steel drill pipe, a motor to turn and lift the drill pipe, and a sturdy mast to grip and support the pipe. Figure 6.6 illustrates an example of this process. A mixture of bentonite clay or other materials is used in combination with water to improve the ability to lift the cuttings out of the borehole. This mixture is called "drilling mud" or just "mud." There are many different kinds of rotary drilling rigs, but they can be summarized into two basic setups: a table drive unit and a top-head drive unit. A table drive drilling rig rotates the pipe using a pipe grip and spinning mechanism near the base of the rig. A top-head drive turns the drilled pipe by way of a motor attached to the upper end of the pipe. In both setups, the drill pipe is also attached to a lifting mechanism that lowers and raises the pipe along the mast. A swivel on top of the pipe is present in both setups, allowing the drilling mud to be pumped down the drill pipe while it is rotating.

Mud rotary well drilling is much faster than using the cable drilling technique, and mud rotary machines are capable of drilling a borehole of up to 60 *cm* or more in diameter. They can achieve depths of up to 60 *m*. In comparison with the cable drilling technique. though, mud rotary rigs are more energy-intensive and they require more fuel per hour to power them. Additional components on this machine beyond a cable drilling rig include a motor to rotate the pipe column, the pipe winch, and the mud pump [11].

2.3.7. AIR ROTARY

The primary difference between the air rotary drilling method and the mud rotary drilling method is that the air technique utilizes compressed air to remove the cuttings rather than drilling mud. A type of "foam"

FIGURE 6.7. An air rotary drilling rig in transport configuration [31]

FIGURE 6.8. A well point for a 2 *in* pipe driven tube well installation [10]

can be added to the air stream in order to improve its effectiveness at the cuttings removal and provide additional stability to the borehole. The mechanical elements of the pipe mechanisms on the mud rotary and air rotary machines are the same. Both styles of machine can come with either a table drive unit or a top-head drive unit. Both require a pipe winch. The air rotary rig utilizes the same type of drill bits as that of a mud rig but also makes use of a down-the-hole hammer drill action. The bit used in air rotary rig operations directs a jet of compressed air to break up rock and drill extremely fast. This type of drilling technique can be setup very quickly, since no mud or cutting mix is utilized, only compressed air. This method is able to drill much faster than other rigs of comparable size and creates less of a mess at the bore site. However, the air compressors utilized by air rotary rigs are generally very large, which adds additional capital cost, potential maintenance needs, and further increased fuel use [11]. An example of a typical air rotary drilling rig is shown in Figure 6.7.

2.3.8. DRIVEN WELLS (TUBE WELLS)

Driven wells require the following four primary components: a well point, well point couplings, lengths of galvanized steel pipe, and a well point drive cap [10]. Each of these components is displayed in Figures 6.8–6.11. Galvanized pipe is required for long-term water system integrity and is commonly available [33]. Driven wells and shallow tube wells are constructed by driving a small diameter pipe into a shallow water-bearing soil profile composed of primarily sand or gravel [12, 32]. Unlike the other well construction techniques mentioned previously, material is not removed but instead is forced aside during the driving process [31]. A screened well point is attached to the bottom of the casing before driving [12]. Couplings are used to connect each section of piping as needed. The drive cap is screwed onto the upper end of the section of pipe that will be driven so as to protect the pipe threads during driving. "The drive and couplings, in addition to being heavier than standard pipe, are designed so that the pipe ends butt together inside the coupling, resulting in most of the driving force being transmitted by the ends of the pipe rather than by the threads" [31]. Driving is done by alternately raising and dropping a weight, which is used as the driving ram. A drive point and manual installation are shown in

FIGURE 6.9. A drive coupling for a 2 *in* driven tube well stack [5]

FIGURE 6.10. A typical unthreaded piece of potable water-safe galvanized 2 *in* steel pipe [33]

FIGURE 6.11. A tube well installation drive cap [5]

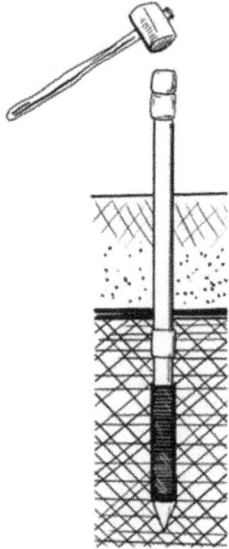

FIGURE 6.12. A cross-section view of a manual installation of a drive tube well installation [11]

FIGURE 6.13. Cross-section view of a device for well driving by a guiding sleeve outside of the well pipe [31]

FIGURE 6.14. Cross-section view of device for well driving by a guiding lineup probe inside of the well pipe [31]

Figure 6.12. In place of only using a hammer to drive, guides can be employed to direct the pipe during driving. This can be done by having a guide either on the outside of the pipe or the inside of the pipe as shown in Figures 6.13 and 6.14, respectively [31]. Prior to the well-driver PUP, driven wells were typically installed manually or using a derrick constructed in place for the specific purpose [10].

The achievable depth to which a tube well can be driven depends on the buildup of friction between the well pipe, the material penetrated, and the transmission of the force of the driver down the length of pipe [31]. It is possible to achieve maximum depths of 25–30 *m*, but hard formations cannot be penetrated [31]. Tube wells are most easily installed in locations where the soil profile is mostly loose sand and the water table is high. Locations close to a river, lake, or stream are especially good [11]. Driven wells are simple and economical to construct. Driven wells are generally not sealed at the surface using grouting material [12], and therefore they may normally lack an adequate sanitary seal [31]. For these reasons, this type of well is more prone to contamination. However, the proper finish with an apron and riser pipe can prevent most common contamination issues [10]. Manually driven wells are not generally able to penetrate more than 5 *m* below the surface [12], but this level has already been surpassed by the mechanized well-driver PUP technology [3].

3. WELL-DRIVER PUP

The PUP vehicle is a development of the Agricultural and Biological Engineering Department at Purdue University under the direction of Dr. John Lumkes. This device is a versatile, useful, and inexpensive tool for the developing world. Researchers at Purdue have modified the basic device to install tube wells [10]. This section will introduce the PUP vehicle, the hydraulic ram attachment, the potential for impact worldwide, the components needed to drive a tube well, and the results from preliminary work. A review of healthy water quality will be provided, along with an estimated cost structure for these types of installations.

FIGURE 6.15. PUP vehicles produced under the direction of Dr. John Lumkes of the Purdue University Agricultural and Biological Engineering Department [10]

3.1. PURDUE UTILITY PLATFORM WITH WELL-DRIVER ATTACHMENT

A hydraulic post driver mated to a PUP vehicle has been designed to mechanize the process of installing driven water wells. This machine has been designated the well-driver PUP [10]. A PUP is a three-wheeled low-cost utility vehicle designed at Purdue University for use in developing countries [3, 34]. These vehicles are typically built in-country with minimal tooling and using only locally sourced materials. The experience base for this vehicle is predominantly in sub-Saharan African countries, mainly Guinea, Cameroon, Nigeria, Uganda, and Kenya [34, 35]. Figure 6.15 shows multiple examples of Purdue student-built PUP vehicles. These vehicles have been used previously for light commercial transportation purposes, including use as a small pickup truck, a taxi, a fire truck, and miscellaneous hauling in areas where normal vehicles are not appropriate due to the severe terrain and lack of road accessibility. These vehicles have even been used as ambulances to get individuals in need of medical services from extremely rural areas to hospitals and clinics. The vehicles have also been used for transporting goods to and from market and as school bus alternatives, garbage collection vehicles, and light utility tractors. Tillage attachments, seeders, harvest heads, water pumps, generators, threshers, and maize grinders powered by the PUP have been designed and tested [3, 15, 16, 34, 36]. These implements have helped improve smallholder farmer access to markets and have improved livelihoods for many of those in sub-Saharan Africa. The well-driver PUP is a further example of alternative use for this versatile vehicle and is displayed in Figure 6.16. The well-driver PUP could potentially reduce dependency on manual labor for driven well installation in developing countries, improve productivity, and keep laborers safer [3, 10, 15].

In locations where the number of drilling rigs is reduced and labor is scarce, the well-driver PUP vehicle could be used as an instrument of economic development, providing microbusiness development opportunities focused on well installation. Installing low-cost driven wells would improve the availability, quality, and accessibility of water in locations lacking sufficient water supplies. In order to be successful, the well-driver PUP operation and components must remain low-cost, easy to maintain, and based on locally accessible materials as much as possible. Although the efficacy of the effort might be diminished, it is intended that operation of this vehicle

FIGURE 6.16. Well-driver PUP vehicle in operational configuration with the hydraulic driving ram erected [37]

should not require formal training in well drilling or geology. In locations of appropriate water table depth, this vehicle could decrease both the wait time to install and the final cost of installed sanitary water wells [10].

3.2. POTENTIAL LOCATIONS OF UTILITY FOR THE WELL-DRIVER PUP IN THE DEVELOPING WORLD

Global estimates indicate that 68% of Earth's freshwater resources are locked up in ice and glaciers, and 30% are found in the ground [38]. Groundwater is the portion of the total precipitation that soaks into the earth's crust and percolates downward into the porous spaces in the soil and rock, where it remains or potentially from where it finds its way out to the surface [3, 10]. Groundwater serves as the world's largest source of freshwater and plays a critical role in meeting the household needs of people around the world [39]. Groundwater is the primary source of the world's drinking water and supplies water for agricultural and industrial activities worldwide [40].

Through testing, Horn demonstrated that the well-driver PUP could achieve depths of up to 7.0 m [3, 10]. Although the vehicle has not yet been demonstrated through formal experimentation, Horn analytically determined that the vehicle should be capable of driving to depths of up to 15 m without significant changes to the apparatus [10]. The depth to the water table varies throughout the world, but significant areas of the world have water within 15 m. Ongoing work is aimed at increasing the proven maximum achievable depth for water wells with this equipment [5].

Prior to the work of Horn [10], the design of the well-driver PUP was a project of several senior capstone teams in the Agricultural and Biological Engineering Department at Purdue University. The first capstone team mounted the post driver to the PUP frame using static three-point hitch lower arms [41]. This design pivots on the rear balls of the lower link arms, thereby allowing the driving ram to be rotated by a hydraulic cylinder serving as the upper link arm. This allows the post driver vehicle to have a more distributed weight during transport operations [42]. This has the effect of moving the mechanical driver from a vertical operational orientation to an inclined transport position.

FIGURE 6.17. Well-driver PUP vehicle in transport configuration with the hydraulic driving ram collapsed into holding cradle [44]

A second capstone team refined the well-driver PUP with additional safety shielding around the hydraulic pump, fixed hydraulic leaks, and made various additional vehicle improvements [43]. A support cradle was added to allow the post driver to stabilize and minimize bouncing during travel. A transmission lockout was also added to force the vehicle's transmission to remain in neutral while operational at high engine speeds. This prevented the vehicle, if bumped by accident during operation, from jumping into gear [43]. Various additional improvements were made to the vehicle by Horn [10] before any initial driving efforts could be carried out. Modifications included rebuilding the engine, improving the efficiency of the hydraulic system, and fabricating outriggers to address the stability and weight distribution issues of the vehicle during well driving operation. Figure 6.17 portrays the well-driver PUP in the transport position. One of the four outriggers designed by Horn [10] is clearly visible on the right rear of the PUP vehicle. The driver support members are also highlighted in the photograph. A third capstone team added well pipe support grips [44], and a fourth team worked on vehicle repairs and supplemental ram weight [45].

3.3. WELL COMPONENTS REQUIRED FOR TUBE WELL DRIVING

The driving action performed manually when installing hand-driven tube wells has been mechanized hydraulically by the well-driver PUP. The components required to install a tube well stack utilizing the well-driver PUP are similar to that of a standard hand-driven well. These components include the well point, the drive couplings, galvanized steel piping, and a well point drive cap. These parts were utilized by Horn [10] during his experimentation and have previously been shown in Figures 6.8–6.11. In addition to these components, Horn fabricated a specialized tube well installation drive sleeve appropriate for driving 2 *in* diameter pipe [10].

The Horn-manufactured well sleeve is shown on the left in Figure 6.18. The driving sleeve shown on the right in Figure 6.18 is used for fence posts and was created by Shaver Manufacturing [10]. "The [well pipe] drive sleeve was constructed of 4–1/2″ × 3/16″ wall-drawn over-mandrel (DOM) steel tube, ½″ steel rod, and ½″ A36 steel plate" [10]. The DOM tubing was specifically chosen because the internal weld bead is ground flush as opposed to a raised weld bead, which would have posed alignment problems and caused unnecessary damage to the well point drive cap. The fabricated well point drive sleeve cleared with an

FIGURE 6.18. A comparison of the Horn 2 *in* tube well installation drive sleeve (left) and the Shaver (Graettinger, Iowa) steel fence post driving sleeve [10]

FIGURE 6.19. Horn 2 *in* tube well installation drive sleeve positioned in hydraulic ram channel [10]

approximated 0.6 *cm* gap around the diameter of a well point drive cap for a 2 *in* diameter pipe. Rub rails made from ½ *in* steel rods were used for the inner steel channel and to position the center of the steel tube sleeve in line with the center of the driver strike plate. A ½ *in* steel plate was welded to the top of the drive sleeve to act as a "cap," serve as a wear plate, and hold the sleeve in the proper position for striking [10]. The drive sleeve used by Horn [10] was designed for installing wells with a 2 *in* diameter pipe and is shown in position within the driving ram in Figure 6.19. The installation of a well with a different pipe diameter would require the fabrication of a new drive sleeve with similar features that matches the new desired well pipe diameter.

3.4. KEY EXPERIENCE GAPS TO BE ADDRESSED IN WELL-DRIVER WELL INSTALLATIONS

Horn [10] hit water at some point during the driving process for the five test well installations that were completed. The results are shown in Table 6.2. These wells were all installed in Montgomery County, Indiana. The recovered water from the submersible pump varied in terms of providing continuous flow, intermittent flow, and no flow [10]. Based on water supply ratings from these test installations, wells #1, #2, #4, and #5 were all deemed "dry" or "intermittent" wells. Well #3 provided a continuous water supply and therefore was developed into a quality water well for testing purposes. This well had a depth of 7.0 *m* and a static water column of 6.1 *m* within the well [10].

The well pump utilized was a Waterra WSP-12 V-3B (Mississauga, Ontario) [46], which moved water at approximately 11.0 *Lpm*. Well #3 was completed, or finished, in accordance with the American Groundwater Trust procedure through the addition of a concrete pad, well surging, and disinfection [47]. The surging

TABLE 6.2. *Summary of the Purdue well-driver PUP experience at tube well driving in Montgomery County, Indiana [3, 10]*

WELL #	WELL DEPTH (M)	STATIC WATER COLUMN IN WELL (M)	CONFINING LAYERS (M)	WATER SUPPLY
1	4	0.9	N/A	N/A
2	3	0.3	N/A	Intermittent
3	7	6.1	N/A	Continuous
4	5.2	N/A	2.7–3.4, 4.6–5.2	N/A
5	7	N/A	N/A	N/A

process removes the fine particles accumulated at the bottom of the well tube near the pump inlet and prevents these from being transferred into the drinking water drawn from the well [29, 48]. Surging also helps pack layers of fine particles around the well water inlet screen, which can then act as a prefilter for large particles. The installation of a driven well of any other size than the 2 *in* diameter by the well-driver PUP would have required the purchase or fabrication of a well point, a drive cap, couplings, and galvanized steel piping to match the diameter of the pipe desired. A larger pump would also have been required to surge and condition the larger-diameter well for finishing.

3.5. WATER QUALITY RESULTS

Water quality samples were collected and submitted to the Montgomery County Health Department for analysis. The water quality parameters evaluated were total coliform and E. coli count. Fecal coliform count was not checked. These tests were done on a present/absent basis. Installation of any new well in Indiana requires that upon receiving a continuous flow rating, a water quality sample must be collected and sent to the perspective county health department for analysis. This was carried out by Horn [10] in accordance with state regulations [29, 48]. The water quality test for well #3 was reported to be absent for total coliform and E. coli count and therefore satisfactory. Since the Montgomery County Health Department deemed the sample satisfactory, the water sample is considered at the time of examination bacteriologically safe based on U.S. EPA standards [3, 10, 29, 48].

3.6. POTENTIAL IMPACT

A high-resolution global-scale groundwater model was developed by de Graaf et al. [39]. A high-resolution global-scale groundwater model was created, highlighting the current water table depth on a global scale through computer model simulations. Most global-scale hydrological models do not include groundwater flow as a component of the model due to the lack of consistent geohydrologic data available on a global scale. This model, run at 6° of resolution, utilized MODFLOW (US Geological Survey, Washington, DC) to construct an equilibrium water table at its natural state. The aquifer schematization and properties used were based on the globally available datasets of lithology and transmissivities, combined with the thickness of an upper unconfined aquifer. The model was initialized using outputs from the land-surface PCRaster Global Water Balance (PCR-GLOBWB) model (Utrecht University, Utrecht, Netherlands), which included net recharge and

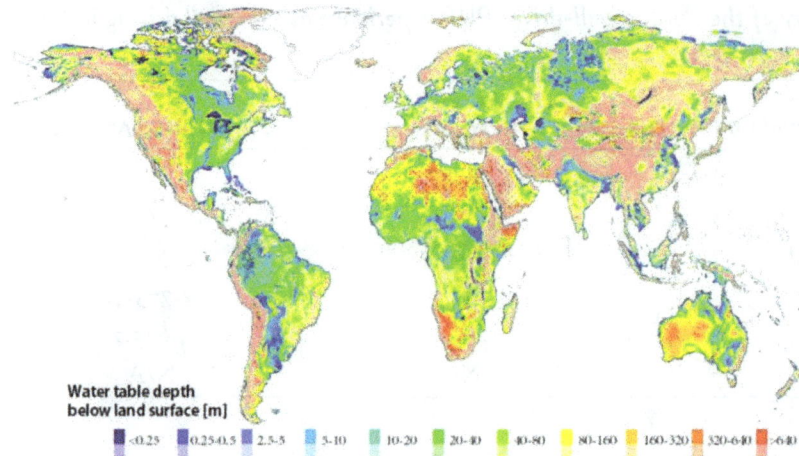

FIGURE 6.20. Output for the estimated water table depth below land surface (*m*) from a geological model [39]. The water table is most shallow in eastern North and South America; central Africa; and central Europe. It is deepest in western North and South America; Saharan and south Africa; southern Europe; the Arabian Peninsula; eastern Asia; and western Australia.

surface water levels. A sensitivity analysis of the various parameter settings was performed and showed that the greatest variation in saturated conductivity had the largest impact on estimates of the groundwater levels. The model validation with observed groundwater levels demonstrated that the predicted levels are reasonably well simulated for many regions of the world, particularly for sediment basins (R^2 = 0.95). These simulated regional-scale groundwater patterns help to provide insight into the availability of groundwater globally [39].

Figure 6.20 provides the expected water table depth below the land surface throughout the world based on the de Graaf et al. simulation [39]. Worldwide, there are many locations where the water table depth is projected to be within the 10–20 *m* range according to this model. Many of these locations are in sub-Saharan Africa, South America, northern India, Asia, and parts of the Asia Pacific Islands. These predictions simply identify potential locations where the well-driver PUP might be utilized if sufficient depth can be demonstrated on a repeatable basis.

According to the United Nations, approximately 14.5% of the world's population is located in sub-Saharan Africa [1]. Earth is projected to hold 9.8 billion people in 2050 and 11.2 billion people by 2100 [49]. The population of sub-Saharan Africa alone is predicted to nearly double by 2050 [1]. Of the 10 largest countries worldwide, Nigeria, which is located in sub-Saharan Africa, is growing the most rapidly [49]. Nigeria is projected to surpass the United States in population and become the third-largest country in the world shortly before 2050 [49]. Clearly, when considering locations where the well-driver PUP could impact the largest number of people, consideration should be given to locations in sub-Saharan Africa.

The International Groundwater Resources Assessment Centre (IGRAC) and the UNESCO International Hydrological Programme have mapped out the transboundary aquifers in Africa [50]. IGRAC, in collaboration with the British Geological Survey and the University College London, has developed maps to quantify the groundwater resources in Africa based on local well data and known geological conditions. Their results for aquifer depth are highlighted in Figure 6.21 and provide a high-resolution look at the depth to water table in Africa. When specifically looking at sub-Saharan Africa, the approximate depth to the groundwater is predominantly less than 25 *m*, followed by areas in the 25–50 *m* range. These data demonstrate the vast number of locations in sub-Saharan Africa where the well-driver PUP could possibly access groundwater. Once fully developed, this technology has great potential to have a substantial positive impact for the people in those regions of the world with little to no water access.

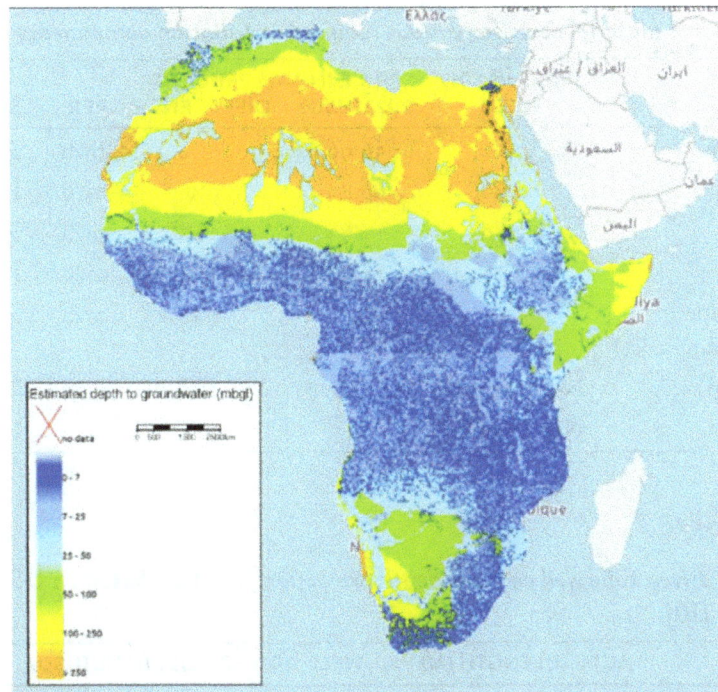

FIGURE 6.21. Estimated depth to groundwater (*mbgl*) and transboundary aquifer of Africa based on geologic and well drilling data [50]. The groundwater is deepest in the Sahara and southwestern portions of the continent. It is shallower in sub-Saharan Africa.

3.7. POTENTIAL PEOPLE SERVED PER WELL ESTIMATES

The economic analysis of a water well installation depends on the number of people the well can serve, the depth of the water being pumped, the daily per capita water required, the duty cycle of the water pump, and the ability of the water-bearing formation to be sustained during pumping [10]. Additionally, the water table drawdown may affect the pumping flow rate over time. The successful water well installed by Horn [10] had a depth of 7 *m* and a pumping rate of approximately 11 *L/min*. If a continuous duty cycle on this well was run for 8 *h* per day, a total of 5400 *L* of water could be pumped [10]. According to the Sphere humanitarian standards for developing countries [4] and the World Health Organization [8], the minimum water requirement is 15 *L* per person per day. Therefore, with a hand pump, the maximum number of people using the Horn #3 well on a per day basis should not exceed 500, and a minimum flow rate of 16.6 *L/min* would be needed [4]. This guideline assumes that the water point is accessible for approximately 8 *h* of the day [4]. Using these developed metrics, Horn's initial successful test well [10] could therefore supply water for between 270 and 360 people if installed in a community setting.

3.8. WELL COST PER DEPTH AND VALUE PROPOSITION ESTIMATES

Horn [10] conducted a cost analysis per well depth ranging from 1 to 35 *m* using the US market prices of the driven well components in 2019. This was updated to reflect current prices of those components in 2022, and the results are shown in Table 6.3. The percent difference between the pricing periods was calculated to determine how much change in the costs have occurred during the intervening years. Even with the recent spike

TABLE 6.3. *Driven tube well material costs for various required installation components [10]*

MATERIALS FOR WELL	PRICE IN 2019 (USD)	PRICE IN 2022 (USD)	PERCENT DIFFERENCE	PER	SOURCE	LENGTH (M)
2″×10″ galvanized steel pipe	37.12	56.96	35%	each	HD	3
2″×36″ well point	60.66	66.89	9%	each	HD	1
2″ pipe coupling	12.86	11.63	−11%	each	HD	0
2″ well point drive cap	17.29	15.99	−8%	each	HD	N/A
2″×5″ galvanized steel pipe section (plus cut/threading cost)	20	20a	0%	each	HD (approx.)	1.5
Submersible water pump	Not reported	224.1	0%	each	Waterra	N/A
1−80 *lb*$_f$ bag of concrete	Not reported	5.87	0%	each	HD	N/A

TABLE 6.4. *An updated driven tube well cost per depth projection for 2022 based on US costs and a comparison against prices in Ghana [10]*

WELL DEPTH (M)	ACTUAL LENGTH (M)	WELL-DRIVER COST IN US (USD)	COST IN GHANA (USD)
1.5	2.4	512	565
3	4	530	687
4.6	5.5	549	809
6.1	7	567	931
7.6	8.5	585	1053
9.1	10.1	604	1175
10.7	11.6	622	1297
12.2	13.1	640	1419
13.7	14.6	659	1541
15.2	16.2	677	1663
16.8	17.7	695	1785
18.3	19.2	714	1907
19.8	20.7	732	2028
21.3	22.3	750	2150
22.9	23.8	769	2272

in steel prices, the net changes have been relatively small. Table 6.4 provides an updated well cost per depth using the current prices in 2022, modeled after Horn's initial calculations. This table also compares tube well costs for installations in Ghana at the same equivalent depth. One meter is the length of the 2 *in* well point, and it was therefore selected as the minimum depth [10]. Economic assumptions included in the analysis were that the well casing extended 1 *m* below the ground surface and that any section of pipe used that was less than 1.5 *m* was considered to have the full cost of a 1.5 *m* section [10]. A maximum depth of 15 *m* was chosen, as recommended by the book *Groundwater and Wells* [14]. The text states that well points driven by hammers massing 113 to 454 kg (1.1–4.5 kN) should be able to reach depths of 15 m or more under favorable situations [14]. The effective weight of the spring-powered driving ram, per the Shaver HD-8® operator's manual, is 1.6 kN, which indicates that a 15 m deep well should be within an acceptable potential cutoff range for the driver without any modifications [10, 51]. This cost per depth table is carried out to below 15 m in the event that such depths can be reached through further development.

If additional modifications can be made to allow the driver to attain still deeper depths, then Table 6.4 could easily be expanded. The average cost of a drilled well in the United States without a well casing is reported to have a range of US\$49–98/m, or potentially up to US\$164/m in tough soil conditions [52]. By averaging the cost per depth from Table 6.4, a driven well within the capabilities of the driver would average about US\$68.05/m in physical material costs, excluding the well-driver PUP vehicle cost and fuel costs.

Based on these calculations, it is possible that a driven well could be cheaper per m, even in a US context, than previously reported by Horn [10]. A tube well, driven by a well-driver PUP, could certainly be a cost-effective alternative to a drilled well in a developing country. The current average cost per depth in Ghana is US\$80/m for a tube well and US\$109/m for a drilled well [53, 54]. These values were first cited by Namara et al. [53] in 2011, and in this work they have been updated to account for the depreciation of the Ghana cedi between 2011 and 2022. The price increases for steel have nearly doubled its total cost since the Horn study [10] in 2019, clearly necessitating a reevaluation of the cost structure for tube well installation. The calculations in Table 6.4 account for pump, concrete, and labor costs. According to the US Bureau of Labor Statistics [55], the average hourly wage for labor in the water, sewage, and other systems industry is \$34.86 US per h [55]. In Ghana, the hourly rate is \$2.76 US per h [56]. It is assumed that six hours of labor would be required for a well installation. Fuel costs are excluded from these calculations due to variation of fuel pricing in both the United States and Ghana. This indicates that using the well-driver PUP to install a driven well in Ghana would be on average a potential savings of 49% compared to the current process. There are numerous locations throughout the world and, more specifically, in sub-Saharan Africa where the depth to groundwater is within the well-driver PUP's potential depth of 15 m [39, 50]. However, this proposition requires further testing to prove that that this depth can be repeatedly achieved, since the Horn 2019 study [10] only demonstrated an experimental depth of 7 m. Furthermore, increasing the achievable driving depth capabilities of the vehicle through design improvements would increase the number of locations across the globe where the vehicle could provide improved access to groundwater.

Over the last eight years, the primary author has had significant professional experience working in Ghana on various international development projects related to agriculture and WASH programming and continues to have access to a variety of networks in the country that are expressing interest in the use of the well-driver PUP once it becomes proven and commercially viable. Therefore, consideration was initially given to the appropriateness of the well-driver PUP if tested and eventually available within Ghana. Located in sub-Saharan Africa, Ghana has a long history of accessing shallow groundwater for the purpose of agricultural irrigation [57]. This experience has predominantly been in the Keta strip in the Volta region. Groundwater has been accessed through various power sources to lift water, including human feet/hand-operated equipment, such as ropes and buckets. Access to many traditional shallow groundwater sources was extremely labor-inefficient, often being via hand-dug wells. Newer systems exist, but they are often out of reach for an individual farmer or household due to the capital costs of acquisition. Currently, there are a total of 34,263 wells recorded in the Keta district. However, these are predominantly used for irrigation purposes. The small depth range of current tube wells in this area is between 6 to 9 m, and water is lifted primarily through small electrically powered pumps located near the tube well. There is increasing potential in this area due to the shallow alluvial depths of the aquifers being less than 20 *m* below the surface throughout the dry season [57].

Ghana's precipitation, surface water, and largely untapped groundwater resources are wholly sufficient to meet most of their projected water needs [58]. Ghana's groundwater resources are predominantly untapped with ample room for scale-up, particularly for agricultural purposes [59]. Ghana's groundwater aquifers range from between 10 and 60 *m* in depth, with well yields rarely exceeding 6 *m³/h*. However, these well yields can be much higher and the depths can be much deeper in areas where limestone is present in the soil profile [60]. As highlighted by de Graaf et al. [39] and IGRAC [50], portions of Ghana's groundwater are accessible within 15 *m* of the surface. In areas where the water table is within Horn's previously demonstrated depth, such as portions of the Keta strip, the potential to benefit individuals already exists at 7 *m*. However, if depths of up to 15 *m* can be achieved, a far greater number of locations could also benefit from this well installation process. As per Table 6.4, a 49% cost savings to install shallow tube wells in Ghana by using the well-driver PUP technology is extremely promising.

4. CONCLUSION

In developing countries, water is not always palatable and available in sufficient quantities. In many locations around the world, people lack sufficient access to water for both drinking and domestic purposes and use unsafe water sources. Water-related diseases pose a major risk to individuals through the consumption and unsafe use of poor water quality sources [4]. This is particularly true in sub-Saharan Africa. People must have equitable and affordable access to safe and sufficient water that is potable and in sufficient quantity for both drinking and domestic purposes. Stored underground water that can be removed by wells is the most likely means to supply this need. In many countries around the globe individuals obtain their drinking water from community wells, so this kind of water access is commonly used for drinking and domestic purposes.

Worldwide, there are many locations where the water table depth is less than 15 *m*, specifically in the 10–20 *m* range. Many of these locations are in sub-Saharan Africa. Ghana is one of the many countries located in sub-Saharan Africa where the well-driver PUP could have a positive impact on the quality of life for those living there. Horn [10] installed a series of test wells, with the deepest being 7.0 *m*. This well received a continuous water rating and was formally completed [10]. The well water quality results analyzed received a satisfactory rating from the Montgomery County health authorities [3, 10, 29, 48]. Horn's initial test well could potentially supply water for between 270 and 360 people if installed in a similar community setting.

This chapter reviewed the three primary types of wells utilized to obtain groundwater resources: dug, drilled, and driven wells [11]. Water wells of these three types can be installed through either manual or powered methods [11]. The well-driver PUP is a low-volume manufactured utility vehicle with a hydraulic post driver mated to it to mechanize tube well installation. The components required to install a driven tube well stack utilizing the well-driver PUP include the well point, the drive couplings, galvanized steel piping, a well point drive cap, and a drive sleeve. The implementation and dissemination of the well-driver PUP technology has the potential to improve access to safe water in developing countries for both drinking and domestic purposes.

5. ACKNOWLEDGMENTS

The authors wish to thank the reviewers and journal editors for their gracious contribution of time and effort to improve this manuscript and ensure its publication. Dr. John Lumkes is recognized for his gracious donation of a PUP vehicle to further this research. Dr. Carol S. Stwalley is thanked for her editorial and proofreading services in preparation of this manuscript. Elvis Kan-uge is thanked for his review and cultural insight. All cited authors who provided permissions for the reuse of their materials are hereby acknowledged and respectfully thanked. This research did not receive any specific grant resources of any kind from agencies in the public, commercial, and not-for-profit sectors. However, the assistance of the Purdue University Agricultural and Biological Engineering Department is gratefully acknowledged for its support over the years with graduate teaching assistantships and faculty salaries. Purdue University is an equal opportunity/equal access employer and service provider.

6. CONFLICT OF INTEREST

The authors declare no conflict of interest.

7. SHALLOW TUBE WELL TECHNOLOGY QUESTIONS

1. Explain the difference between surface water collection and groundwater access. List five potential contaminates that are filtered through soil infiltration.
2. What are the three categories of human water need for basic survival? What is the minimum daily human need for ingestive consumption?
3. All wells require isolation from direct infiltration of water from the surface. How do drilled and driven wells protect the underground aquifer from surface contamination?
4. Explain the importance of the drive sleeve on top of the tube well stack during the installation process.
5. Describe the advantages of tube well installations compared to drilled water wells in the developing world. How does the cost of labor affect the calculus of planning a water well? How does the cost of galvanized steel pipe and other supplies affect the proposed installation?

8. REFERENCES

[1] United Nations Department of Economic and Social Affairs, Population Division. World Population Prospects 2019: Highlights. UN, New York. 2019. [Online]. Available: https://www.population.un.org/wpp/Publications/Files/wpp2019_10KeyFindings.pdf [last accessed December 15, 2020].

[2] C. Sweetman and L. Medland, "Introduction: Gender and water, sanitation, and hygiene," *Gender and Development*, vol. 25, no. 2, pp. 153–166, 2017. https://doi: 10.1080/13552074.2017.1349867.

[3] Z. J. Horn and R. M. Stwalley III, "Design and testing of a mechanized tube well installer," *Groundwater for Sustainable Development*, vol. 11, 100442, 2020. https://doi: 10.1016/j.gsd.2020.100442.

[4] Sphere, *The Sphere Handbook: Humanitarian charter and minimum standards in humanitarian response*, Geneva: Sphere. 2018. [Online]. Available: https://www.spherestandards.org/handbook-2018/ [Last accessed December 15, 2020].

[5] G. L. Baldwin and R. M. Stwalley III, "Improving the driving capabilities of a well-driver PUP (Purdue Utility Project) to install low-cost driven water wells, in *ASABE Annual International Meeting, Houston*, 2022. https://doi: 10.13031/aim.202200203.

[6] B. Nolan, K. Hitt, and B. Ruddy, "Probability of nitrate contamination of recently recharged ground waters in the conterminous United States," 2002, [Online]. Available: https://www.water.usgs.gov/nawqa/nutrients/pubs/est_v36_n010/ [Last accessed December 15, 2020].

[7] P. J. Weyer, J. R. Cerhan, B. C. Kross, G. R. Hallberg, J. Kantamneni, G. Breuer, et al. "Municipal drinking water nitrate level and cancer risk in older women: The Iowa Women's Health Study," *Epidemiology*, vol. 12, no. 3, pp. 327–338, 2001. https://doi: 10.1097/00001648–2,001,050,000–00013.

[8] World Health Organization, "What is the minimum quantity of water needed?," 2016. [Online]. Available: https://www.who.int/water_sanitation_health/emergencies/qa/emergencies_qa5/en/. [Last accessed January 1, 2019].

[9] UNICEF, "Collecting water is often a colossal waste of time for women and girls," 2021. [Online]. Available: https://www.unicef.org/press-releases/unicef-colelcting-water-often-colossal-waste-time-women-and-girls. [Last accessed December 16, 2021].

[10] Z. J. Horn, "Prototyping a well-driver PUP (Purdue Utility Project) to install low-cost driven water wells," Master's thesis, Purdue University.

[11] F. Proby, "Shallow well drilling," 2013. [Online]. Available: https://www.lifewater.org/wp-content/uploads/2018/10/lifewater-Well-Drilling-Manual.pdf. [Last accessed October 11, 2021].

[12] Wellowner. "Types of wells," 2015. [Online]. [Last accessed November 17, 2019].

[13] J. Carroll, personal communication. July 14, 2019.

[14] F. G. Driscoll, *Groundwater and Wells*. Saint Paul, MN: Johnson, 1986, LoC 85–63,577.

[15] Z. J. Horn and R. M. Stwalley, "Well-driver PUP," in *ASABE Annual International Meeting—Detroit*, 2018. https://doi: 10.13031/aim.201801196.

[16] Z. J. Horn and R. M. Stwalley III, "A low-cost mechanized tube well installer," in *ASABE 2019 Annual International Meeting—Boston*, 2019. https://doi: 10.13031/aim.201901319.

[17] World Health Organization, "UNICEF progress on sanitation and drinking water—2013 update," WHO Press, Geneva, 2013. [Online]. Available: https://www.data.unicef.org/resources/progress-on-sanitation-and-drinking-water-2013-update/ [Last accessed December 15, 2020].

[18] Millennium Challenge Corporation, "Measuring results of the Ghana water and sanitation sub-activity," 2017. [Online]. Available: https://www.mcc.gov/resources/pub-pdf/report-ghana-closed-compact [Last accessed November 17, 2019].

[19] J. P. Graham, M. Hirai, and S. S. Kim, "An analysis of water collection labor among women and children in 24 sub-Saharan African countries," *PLoS One*, vol. 11, no. 6, e0155981, 2016. https://doi: 10.1371/journal.pone.0155981.

[20] M. Mekonnen and A. Y. Hoekstra AY, "Four billion people facing severe water scarcity," *Science Advances*, vol. 2, no. 2, e1500323, 2016. https://doi: 10.1126/aciadv.1500323.

[21] E. Stevenson, A. Ambelu, B. Caruso, Y. Tesfaye, and M. Freeman, "Community water improvement, household water insecurity, and women's psychological distress: An intervention and control study in Ethiopia," *PLoS One*, vol. 11, no. 2, pp. 1–13, 2016. https://doi: 10.1371/journal.pone.0153432.

[22] E. Bisung and S. J. Elliott, "Psychosocial impacts of the lack of access to water and sanitation in low- and middle-income countries: A scoping review," *Journal of Water and Health*, vol. 15, no. 1, pp. 17–30, 2016. https://doi: 10.2166/wh.2016.158.

[23] P. Routray, B. Torondel, T. Clasen, and W. P. Schmidt, "Women's role in sanitation decision making in rural coastal Odisha, India," *PLoS One*, vol. 12, no. 5, e0178042, 2017. https://doi: 10.1371/journal.pone.0178042.

[24] A. M. Mahama, K. A. Anaman, and I. Osei-Akoto,. "Factors influencing householders' access to improved water in low income urban areas of Accra, Ghana," *Journal of Water and Health*, vol. 12, no. 2, pp. 318–331. https://doi: 10.2166/wh2014.149.

[25] E. Adams, A. Boateng, and G. Amoyaw. "Socioeconomic and demographic predictors of potable water and sanitation access in Ghana," *Social Indicators Research*, vol. 126, no. 2, pp. 673–687, 2016. https://doi: 10.1007/s11205-015-0912-y.

[26] S. A. Sheuya. "Improving the health and lives of people living in slums." *Annals of the New York Academy of Sciences*, vol. 1136, no. 1, pp. 298–306, 2008. https://doi: 10.1196/annals.1425.003.

[27] World Water Assessment Program, "The United Nations world water development report 2015: Water for a sustainable world, facts and figures," United Nations, New York. 2015. [Online]. Available: https://www.unwater.org/publications/un-world-water-development-report-2015 [Last accessed December 15, 2020].

[28] C. Leahy, K. Winterford, T. Nghiem, J. Keheller, L. Leong, and J. Willetts. "Transforming gender relations through water, sanitation, and hygiene programming and monitoring in Vietnam," *Gender and Development*, vol. 25, no. 2, pp. 283–251, 2017. https://doi: 10.1080/13552074.2017.1331530.

[29] Professions and Occupations, "Water Well Drilling," Indiana Code § 25-39-3. 2011.

[30] H2O for Life, "Borehole well," 2019. [Online]. Available: https://www.h2oforlifeschools.org/page/borehole-well. [Last accessed November 17, 2019].

[31] R. G Koegel. "Small diameter wells," 1985. [Online]. Available: https://www.fao.org/3/X5567E/x5567e00.htm#Contents. [Last accessed December 15, 2021].

[32] US Army, *US Army Field Manual: Well Drilling Operations* (FM5–484). US Department of Defense, Washington, D.C., 1994.

[33] Home Depot, "Southland 3/4 in x 10 ft. galvanized steel pipe-564–1200HC.," 2021. [Online]. Available: https://www.homedepot.com/p/Southland-3-4-in-x-10-ft-Galvanized-Steel-Pipe-564-1200HC/100534625. [Last accessed October 27, 2021].

[34] J. P. Robison. "Transportation and power solutions for Africa: The assessment and optimization of the Purdue Utility Platform," Master's thesis, Purdue University, 2016.

[35] Purdue College of Engineering, "Purdue Utility Project," Purdue University, 2019. [Online]. Available: https://www.engineering.purdue.edu/pup/ [Last accessed December 15, 2020].

[36] D. D. Wilson, "Investigation of an affordable multigrain thresher for smallholder farmers in sub-Saharan Africa," Master's thesis, Purdue University, 2016.

[37] G. L. Baldwin and R. M. Stwalley III, "Opportunities for the scale-up of irrigation systems in Ghana, West Africa," *Sustainability*, vol. 14, no. 8716, 2022. https://doi: 10.3390/su14148716.

[38] United States Geological Service, "How much water is there on, in, and above the Earth?," 2016. [Online]. Available: https://water.usgs.gov/edu/earthhowmuch.html. [Last accessed September 30, 2021].

[39] I. E. M. de Graff, E. H. Sutanudjada, L. P. H. van Beek, and M. F. P. Biekens, "A high-resolution global-scale groundwater model." *Hydrology and Earth System Sciences*, vol. 19, no. 2, pp. 823–837, 2015. https://doi: 10.5194/hess-19-823-2015.

[40] Y. Wada, D. Wisser, and M. F. P. Bierkens, "Global modeling of withdrawal, allocation, and consumptive use of surface water and groundwater resources." *Earth Systems Dynamics*, vol. 5, pp. 15–40. https://doi: 10.5194/esd-5-15-2014.

[41] C. Zeller and W. Vaughn, "Well driver," Capstone report, Purdue University, 2016.

[42] S. Zhao, "Well driver PUP vehicle weight distribution during transport and operation," Master's thesis, Purdue University, 2022.

[43] Y. Chen, J. Jarufe, and C. Yun, "Renovations to the Purdue utility well driver," Capstone report, Purdue University, 2016.

[44] J. Adams, T. Hampston, and H. Wilson. "Improvements to the Purdue Utility Platform well-driver," Capstone report, Purdue University, 2019.

[45] M. Fidler, T. McPheron, and R. Soloman. "PUP well driver," Purdue University, 2022.

[46] Waterra, Submersible groundwater sampling pumps: 40, 60, or 90 foot lift," 2022. [Online]. Available: https://waterra.com/product/submersible-groundwater-sampling-pump/. [Last accessed November 19, 2022].

[47] American Ground Water Trus. *agwt.org*, 2012. [Online]. Available: https://www.agwt.org/content/water-well-disinfection-procedure. [Last accessed December 15, 2021].

[48] Water Well Drillers and Well Water Pump Installers, 312 Indiana Administrative Code § 13-1-13-12. 2015.

[49] A. Held, "U.N. says world population will reach 9.8 billion by 2050," National Public Radio, 2017. [Online]. Available: https://www.npr.org/sections/thetwo-way/2017/06/22/533935054/u-n-says-world-s-population-will-reach-9-8-billion-by-2050 [Last accessed December 15, 2020].

[50] IGRAC, "Africa groundwater portal." 2017. [Online]. Available: https://www.un-igrac.org/special-project /africa-groundwater-portal. [Last accessed December 15, 2021].

[51] Shaver Manufacturing, "Operator's manual for hydraulic post driver model HD-8 & HD-8-H," 2009. [Online]. Available: http://www.shavermfg.com/media/uploads/HD8-Operator-Manual.pdf [Last accessed December 15, 2020].

[52] Home Advisor, "How much does it cost to drill or dig a well?," 2021. [Online]. Available: https://www.homeadvisor. com/cost/landscape/drill-a-well/. [Last accessed October 27, 2021].

[53] R. E. Namara, L. Horowitz, B. Nyamadi, and B. Barry, "Irrigation developments in Ghana: Past experiences, emerging opportunities, and future directions," International Food Policy Research Institute, 2011. [Online]. Available: https://www. reliefweb.int/report/ghana/irrigation-development-ghana-past-experiences-emerging-opportunities-and-future#:~:- text=Ghana%20is%20endowed%20with%20sufficient,%3B%20Agodzo%20and%20Bobobee%201,994 [Last accessed December 15, 2020].

[54] E. Kan-uge, Personal communication, May, 20 2022.

[55] US Bureau of Labor Statistics, "Occupational employment and wage statistics," 2022. [Online]. Available: https:// www.bls.gov/oes/current/oes537072.htm#(2). [Last accessed November 19, 2022].

[56] K. Yawson, Personal communication, November, 7, 2022.

[57] R. E. Namara, L. Horowitz, S. Kolavalli, G. Kranjac-Berisavljevic, B. N. Dawuni, B. Barry, et al. "Typology of Irrigation Systems in Ghana," International Water Management Institute Working Paper 142, 2010. https://doi: 10.5337/2011.200.

[58] G. L. Baldwin and R. M. Stwalley III, "A review of freshwater programming and access options in Ghana, West Africa," African Journal of Water Conservation and Sustainability, vol. 10, no. 4, pp. 1–19, 2022. LoC: 2375–0936.

[59] G. L. Baldwin and R. M. Stwalley III. "An economic analysis: The scale-up of irrigation systems in Ghana, West Africa," in ASABE Annual International Meeting—Pasadena, 2021. https://doi: 10.13031/aim.212100009.

[60] Ghanaian Water Resources Commission, "Water use," 2020. [Online]. Available: https://wrc-gh.org /water-resources-management-and-governance/water-use/. [Last Accessed December 11, 2020].

CONCLUSION

The United Nations (UN) classifies freshwater as a necessity for all human beings, but clearly providing access to people in the developing world is not simple. Drilled deep wells are unquestionably the gold standard in groundwater sourcing for people, domestic animals, and agricultural irrigation. However, the lack of available equipment and the expense of the installations limit the progress toward the UN goal. Mechanization of tube well installation represents a potential intermediate means to provide potable water and beginning the economic development of remote and rural areas currently lacking infrastructure and the economic means to install water wells. The peer-reviewed papers compiled in this collection should allow the construction of a mechanized tube well installer. The use of the Purdue Utility Platform vehicle for operations was an expedient choice for the research and should not be considered a requirement. In fact, the researchers would strongly recommend the use of a modest utility tractor as the base vehicle. This would eliminate the need for the leveling stilts and the transport mechanism on the unit, allow for the transport of supplies on a wagon behind the tractor, and make possible the clearing of ground for installation with a bucket loader or blade. If properly planned, a tube well installation business in the developing world could be equipped and provided with an initial round of supplies for under US$50,000. One machine installing three tube wells per week could likely supply freshwater to around 50,000 additional people per year. A fleet of these machines could make significant progress toward meeting the UN goal of freshwater for all within a decade.

COMBINED REFERENCES FROM THE INCLUDED WORKS

Adams, E., Boateng, A., & Amoyaw, G. (2016). Socioeconomic and demographic predictors of potable water and sanitation access in Ghana. *Social Indicators Research* 126(2), 673–687. https://doi: 10.1007/s11205-015-0912-y.

Adams, J., Hampston, T., & Wilson, H. (2019). *Improvements to the Purdue Utility Platform well-driver*. ABE 485 Capstone Experience Final Report, Purdue University, Agricultural & Biological Engineering.

American Ground Water Trust. (2012). Retrieved December 15, 2021, from Water Well Disinfection Procedure. https://www.agwt.org/content/water-well-disinfection-procedure.

American Society for Testing and Materials. (2019). *Standard test materials for laboratory determination of water (moisture) content of soil and rock by mass (standard no. D2216–98)*. West Conshohocken, PA: ASTM. https://doi: 10.1520/D2216–98.

AquaScience. (2017). *Request a free water well test kit*. Retrieved November 1, 2017, from Welcome to AquaScience: https://aquascience.net/free-water-test?srsltid=AfmBOopGg1XE_42CZTdt_NEMSoohpQBzf1JONzyOT-FmoQvTjZxoKqPp8.

Baldwin, G. L. (2019). *Development of design criteria and options for promoting lake restoration of Lake Bosomtwe and improved livelihoods for small-holder farmers near Lake Bosomtwe-Ghana, West Africa* [Master's thesis]. Purdue University.

Baldwin, G. L., & Stwalley, R. M., III. (2018). *An agricultural extension demonstration farm template & community development project* [Paper presentation]. ASABE 2018 AIM—Detroit. 10.13031/aim.201800693.

Baldwin, G. L., & Stwalley, R. M., III. (2019). *Analysis of market assessment survey to help promote lake restoration of Lake Bosomtwe and increased livelihoods for small-holder farmers* [Paper presentation]. ASABE 2019 AIM—Boston. https://doi: 10.13031/aim.201901379.

Baldwin, G. L., & Stwalley, R. M., III. (2020). *Promoting restoration of Lake Bosomtwe through spatial analysis of existing water, sanitation, and hygiene (WASH) sources in Ghana, West Africa* [Paper presentation]. ASABE 2020 AIM—Pasadena. https://doi: 10.13031/aim.202000589.

Baldwin, G. L., & Stwalley, R. M., III. (2021). *An economic analysis: The scale-up of irrigation systems in Ghana, West Africa* [Paper presentation]. ASABE Annual International Meeting—Pasadena. https://doi: 10.13031/aim.212100009.

Baldwin, G. L., & Stwalley, R. M., III. (2022). A review of freshwater programming and access options in Ghana, West Africa. *African Journal of Water Conservation and Sustainability, 10*(4), 1–19. ISSN 2375–0936.

Baldwin, G. L., & Stwalley, R. M., III. (2022). *Improving the driving capabilities of a well-driver PUP (Purdue Utility Project) to install low-cost driven water wells* [Paper presentation]. ASABE Annual International Meeting—Houston. https://doi: 10.13031/aim.202200203.

Baldwin, G. L., & Stwalley, R. M., III. (2022). Opportunities for the scale-up of irrigation systems in Ghana, West Africa. *Sustainability, 14*, 8716. https://doi: 10.3390/su14148716.

Baldwin Kan-uge, G. L. (2023). *Improvements to the driving capabilities of a well-driver PUP (Purdue Utility Project) to install low-cost driven water wells* [PhD dissertation]. Purdue University.

Baldwin Kan-uge, G. L., McPheron, T. J., Horn, Z. J., & Stwalley, R. M., III. (2023). Cost-effective, sanitary shallow water wells for agriculture and small communities using mechanized tube well installation. In J. Tarhouni, *Groundwater—New advances and challenges*. Intech Open. https://doi: 10.5772/intechopen.109576.

Baldwin Kan-uge, G. L., McPheron, T. J., & Stwalley, R. M., III. (2023). *Design and development of a preliminary regression model to determine the driving capabilities of the well-driver PUP* [Paper presentation]. ASABE Annual International Meeting—Omaha. https://doi: 10.13031/aim.202300014.

Bisung, E., & Elliott, S. J. (2017). Psychosocial impacts of the lack of access to water and sanitation in low- and middle-income countries: A scoping review. *Journal of Water and Health, 15*(1), 17–30. https://doi: 10.2166/wh.2016.158.

de Graff, I. E., Sutanudjada, E. H., van Beek, L. P., & Biekens, M. F. (2015). A high-resolution global-scale groundwater model. *Hydrology and Earth System Sciences, 19*(2), 823–837. https://doi: 10.5194/hess-19-823-2015.

Driscoll, F. G. (1986). *Groundwater and wells.* Johnson. LoC 85–63577.

Environmental Literacy Council. (2015). *Water in developing countries.* Retrieved November 1, 2017, from Environmental Literacy Council. http://www.enviroliteracy.org/article.php/1400.html (incorrectly identified in original paper as *Water in Developing Countries* 2016.

Fidler, M. D., McPheron, T. J., & Soloman, R. W. (2022). *Improvements to the well-driver PUP: A final report* [Capstone project]. Purdue University [an unpublished capstone project].

Geschwinder, L. F., & Troemner, M. (2016). Notes on the AISC 360-16 provisions for slender compression elements in compression members. *Engineering Journal, 53*(3), 137–146. https://doi: 10.62913/engj.v53i3.1102 (incorrectly identified in original paper as Matthew 2016).

Ghanaian Water Resources Commission. (2020). *Water use.* Retrieved December 11, 2020, from Water Resources Management & Governance. https://wrc-gh.org/water-resources-management-and-governance/water-use/.

Graham, J. P., Hirai, M., & Kim, S. S. (2016). An analysis of water collection labor among women and children in 24 sub-Saharan African countries. *PLoS ONE, 11*(6), e0155981. https://doi: 10.1371/journal.pone.0155981.

Groundwater Foundation. (2019). https://www.groundwater.org. Retrieved October 15, 2023. from get-informed /basics/groundwater.html.

Hanushek, E. A., & Wossmann, L. (2007). *Education quality and economic growth.* Retrieved November 1, 2018, from WorldBank.org. https://documents1.worldbank.org/curated/en/885051468141302837/pdf/395110Edu1Quality1 Economic1Growth.pdf.

Held, A. (2017). *U.N. says world population will reach 9.8 billion by 2050.* In the Two Way [Amy Held Blog]. National Public Radio.

Home Advisor. (2021). *How much does it cost to drill or dig a well?* Retrieved October 27, 2021, from https://www.homeadvisor.com/cost/landscape/drill-a-well/.

Horn, Z. J. (2019). *Prototyping a well-driver PUP (Purdue Utility Project) to install low-cost driven water wells* [Master's thesis]. Purdue University. https://doi: 10.25394/PGS.8038991.v1.

Horn, Z. J., & Stwalley, R. M., III. (2018). *Well-driver PUP* [Paper presentation]. ASABE Annual International Meeting—Detroit. ASABE. https://doi: 10.13031/aim.201801196.

Horn, Z. J., & Stwalley, R. M., III. (2019). *A low-cost mechanized tube well installer* [Paper presentation]. ASABE 2019 AIM—Boston. https://doi: 10.13031/aim.201901319.

Horn, Z. J., & Stwalley, R. M., III. (2020). Design and testing of a mechanized tube well installer. *Groundwater for Sustainable Development, 11*, 100442. https://doi: 10.1016/j.gsd.2020.100442.

IGRAC. (2017). *Africa groundwater portal.* Retrieved December 15, 2021, from International Groundwater Resources Assessment Centre of the United Nations. https://www.un-igrac.org/special-project/africa-groundwater-portal.

Indiana Administrative Code. (2015). *Article 13: Water well drills and water well pump installers.* Retrieved December 15, 2021, from https://www.in.gov/health/eph/files/312-IAC-13.pdf

Indiana Code (Statutes). (1987). *Article 39: Water Well drilling and pump installer contractors.*

Indiana General Assembly Code. (2010, 1987). *Title 25. Article 39: Water well drilling contractors.* Retrieved December 15, 2021, from https://www.iga.in.gov/legislative/laws/2020/ic/titles/025#25-39.

International Groundwater Resources Assessment Centre. (2017). *African groundwater portal.* Retrieved April 30, 2022, from un-igrac.org: https://www.un-igrac.org/special-project/africa-groundwater-portal.

Kansas Geological Survey. (2001). *KGS—Reno County geohydrology—Ground water recovery.* (2001). Retrieved May 17, 2018, from http://www.kgs.ku.edu/General/Geology/Reno/gwo3.html.

Koegel, R. G. (1985). *Small diameter wells.* Retrieved December 15, 2021, from Food and Agriculture Organization of the United Nations. https://www.fao.org/3/X5567E/x5567e00.htm#Contents.

Krutz, G. W., Schueller, J. K., & Claar, P. W., II. (1994). *Machine design for mobile and industrial applications*. Society of Automotive Engineers, Inc. ISBN 1-56091-389-4.

Leahy, C., Winterford, K., Nghiem, T., Keheller, J., Leong, L., & Willetts, J. (2017). Transforming gender relations through water, sanitation, and hygiene programming and monitoring in Vietnam. *Gender and Development, 25*(2), 283–251. https://doi: 10.1080/13552074.2017.1331530.

Mahama, A. M., Anaman, K. A., & Osei-Akoto, I. (2014). Factors influencing householders' access to improved water in low income urban areas of Accra, Ghana. *Journal of Water and Health, 12*(2), 318–331. https://doi: 10.2166/wh2014.149.

Matthew, T. (2016). *C5.1 Euler's buckling formula*. Retrieved May 12, 2018, from http://www.engineeringcorecourses.com/solidmechanics2/C5-buckling/C5.1-eulers-buckling-formula/theory/.

Mekonnen, M., & Hoekstra, A. Y. (2016). Four billion people facing severe water scarcity. *Science Advances, 2*(2), e1500323. https://doi: 10.1126/aciadv.1500323.

Millennium Challenge Corporation. (2017). *Measuring results of the Ghana water and sanitation sub-activity*. Retrieved November 17, 2019, from mcc.gov/resources/doc/summary-measuring-results-ghana-water-sanitation-sub-activity: https://www.mcc.gov/resources/pub-pdf/report-ghana-closed-compact.

Namara, R. E., Horowitz, L., Kolavalli, S., Kranjac-Berisavljevic, G., Dawuni, B. N., Barry, B., & Giordano, M. (2010). *Typology of irrigation systems in Ghana*. International Water Management Institute (IWMI). https://doi: 10.5337/2011.200.

Namara, R. E., Horowitz, L., Nyamadi, B., & Barry, B. (2011). *Irrigation developments in Ghana: Past experiences, emerging opportunities, and future directions*. International Food Policy Research Institute (IFPRI). Retrieved December 15, 2020, from https://www.reliefweb.int/report/ghana/irrigation-development-ghana-past-experiences-emerging-opportunities-and-future#:~:text=Ghana%20is%20endowed%20with%20sufficient,%3B%20Agodzo%20and%20Bobobee%201994.

National Academy of Engineering. (2021). *NAE grand challenges for engineering*. Retrieved February 12, 2021, from http://www.engineeringchallenges.org/challenges.aspx.

Nolan, B., Hitt, K., & Ruddy, B. (2002). *Probability of nitrate contamination of recently recharged ground waters in the conterminous United States*. Retrieved December 15, 2020, from usgs.org: https://www.water.usgs.gov/nawqa/nutrients/pubs/est_v36_no10/.

Proby, F. (2013). *Shallow well drilling*. Retrieved October 11, 2021, https://www.lifewater.org/wp-content/uploads/2018/10/lifewater-Well-Drilling-Manual.pdf.

Purdue University. (2018). *Purdue Utility Project (PUP)*. Retrieved April 30, 2022, from https://engineering.purdue.edu/pup/where/.

Purdue College of Engineering. (2019). *Purdue Utility Project*. Purdue University. Retrieved December 15, 2020, from https://www.engineering.purdue.edu/pup/.

Routray, P., Torondel, B., Clasen, T., & Schmidt, W. P. (2017). Women's role in sanitation decision making in rural coastal Odisha, India. *PLoS ONE. 12*(5), e0178042. https://doi: 10.1371/journal.pone.0178042.

Shaver Manufacturing. (2009). *Operator's manual: Hydraulic post driver model HD-8 & HD-8-H*. Retrieved December 15, 2020, from http://www.shavermfg.com/media/uploads/HD8-Operator-Manual.pdf.

Sheuya, S. A. (2008). Improving the health and lives of people living in slums. *Annals of the New York Academy of Sciences, 1136*(1), 298–306. https://doi: 10.1196/annals.1425.003.

SimpleSolarHomes. (2013, March 20). *How to install a driven sand point well*. Retrieved March 20, 2018 from https://www.youtube.com/watch?v=I-9g6iZGkoY.

Sphere. (2018). *The Sphere handbook: Humanitarian charter and minimum standards in humanitarian response*. Retrieved December 15, 2020, from https://www.spherestandards.org/handbook-2018/.

Stevenson, E., Ambelu, A., Caruso, B., Tesfaye, Y., & Freeman, M. (2016). Community water improvement, household water insecurity, and women's psychological distress: An intervention and control study in Ethiopia. *PLos ONE, 11*(2), 1–13. https://doi: 10.1371/journal.pone.0153432.

Sweetman, C., & Medland, L. (2017). Introduction: Gender and water, sanitation, and hygiene. *Gender and Development, 25*(2), 153–166. https://doi: 10.1080/13552074.2017.1349867.

UNICEF. (2021). *Collecting water is often a colossal waste of time for women and girls.* Retrieved December 16, 2021, https://www.unicef.org/press-releases/unicef-colelcting-water-often-colossal-waste-time-women-and-girls.

United Nations. (2003). *Meetings coverage and press releases.* Retrieved November 1, 2018, from https://press.un.org/en/2003/sgsm8707.doc.htm.

United Nations Department of Economic and Social Affairs, Population Division. (2019). *World population prospects 2019: Highlights.* United Nations.

United States Geological Service. (2016). *How much water is there on, in, and above the Earth?* Retrieved September 30, 2021, from https://water.usgs.gov/edu/earthhowmuch.html

US Army. (1994). *US Army field manual: Well drilling operations (FM5–484).*: US Department of Defense. ISBN: 979–8746555519.

US Bureau of Labor Statistics. (2022). *Occupational employment and wage statistics.* Retrieved November 19, 2022, from https://www.bls.gov/oes/current/oes537072.htm#(2).

Wada, Y., Wisser, D., & Bierkens, M. F. (2014). Global modeling of withdrawal, allocation, and consumptive use of surface water and groundwater resources. *Earth Systems Dynamics, 5,* 15–40. https://doi: 10.5194/esd-5-15-2014.

Waterra. (2022). *Submesible groundwater sampling pumps: 40, 60, or 90 foot lift.* Retrieved November 19, 2022, from https://waterra.com/product/submersible-groundwater-sampling-pump/

Waterra Pumps Limited. (2018). *Application: Surging.* Retrieved October 12, 2018, from http:www.waterra.com/pages/applications/BodyMovin/surging.html

Wellowner. (2015). *Types of wells.* Retrieved November 17, 2019, from wellowner.org/basics/types-of-wells

Weyer, P. J., Cerhan, J. R., Kross, B. C., Hallberg, G. R., Kantamneni, J., Breuer, G., Lynch, C. F. (2001). Municiple drinking water nitrate level and cancer risk in older women: The Iowa Women's Health Study. *Epidemiology, 12*(3), 327–338. https://doi: 10.1097/00001648-2001050000-00013.

World Health Organization. (2001). *Water for health—Taking charge.* Retrieved October 12, 2018, from www.who.int/water_sanitation_health/takingcharge.html.

World Health Organization. (2013). *UNICEF progress on sanitation and drinking water—2013 update.* Retrieved December 15, 2020, from https://www.data.unicef.org/resources/progress-on-sanitation-and-drinking-water-2013-update/.

World Health Organization. (2016). *What is the minimum quantity of water needed?* Retrieved January 1, 2019, from https://www.who.int/water_sanitation_health/emergencies/qa/emergencies_qa5/en/.

World Water Assessment Program. (2015). *The United Nations world water development report 2015: Water for a sustainable world, facts and figures.* United Nations. Retrieved December 15, 2020, from https://www.unwater.org/publications/un-world-water-development-report-2015.

Wright, F. B. (1977). *Rural water supply and sanitation* (3rd ed.). Krieger Publishing. ISBN: 978–0882753348.

ABOUT THE AUTHORS

ROBERT M. STWALLEY III is an associate clinical professor at Purdue and teaches crop production equipment, power units and power trains, the design of off-road vehicles, computer numerical control (CNC) machining, and machine design in the Agricultural and Biological Engineering Department.

GRACE L. BALDWIN KAN-UGE is an adjunct faculty member at the University of Ghana. Her master's and PhD research at Purdue University focused on improving the lives of citizens living in the developing world. She is currently the chief engineer for Global Resource Connections, Inc., a nonprofit organization dedicated to water resources, food security, health, and poverty reduction in Ghana.

ROGER TORMOEHLEN is a professor in the Agricultural and Biological Engineering Department and the former head of the Department of Agricultural Sciences Education and Communication. He teaches safety in agriculture in the Agricultural and Biological Engineering program and is a highly regarded agricultural sciences educator and researcher.